FOCUS ON CALENDAR AND TIME

K-6 CONTINUITY MATHEMATICS

10 9 8 7 6 5 4 3 2 1

Coursework is available at special quantity discounts to use as premiums and sales promotions within corporate or private training programs. To obtain information or inquire about availability please write to Director, PO Box 1, Hollidaysburg, PA 16648.

FOCUS ON CALENDAR AND TIME

K-6 CONTINUITY MATHEMATICS

Note to Teaching Helper:

The first years of school are transitional years taking a child from a play approach to an academic approach to learning. As you move through the foundations of mathematics, the instruction will become more structured and contain many new concepts and facts for your student to master. It is important that your mathematics lessons encompass a variety of delivery methods. The best programs will encompass both practical worksheets and hands on activities that encourage a deeper understanding of the ideas behind the computations necessary for success in mathematics.

The worksheets included in the skill mastery textbook should serve as an overall guide to the mathematics program and a tool to help you determine your student's mastery of each necessary math building block. You should encourage hands on, creative approach to mathematics as a method of fixing the necessary concepts in your students mind and encouraging a true understanding of the processes of mathematics.

Many learners enjoy using concrete objects or tools to help them master mathematics concepts. You should work with your student to decide what concrete tools work best with their learning style. You might choose any group of pre-made math manipulative, common household items, tally marks, or almost any item that illustrates a specific sum of objects to help your student on the path to a true understanding of the building blocks of mathematics.

The important factor to remember as you move through the each level of the mathematics continuity program is that mastery of each new skill exceeds the need for completion of the entire program during a pre-set term. Statistics have shown that a child will succeed at mathematics only if they gain comprehension of each new concept. You should use the included guide and worksheets to assess your child's progress and only move on to the next segment once you are certain that mastery has been achieved.

The proven path to mastery in mathematics is practice, practice, practice. This does not mean that you should present mathematics in the form of mindless repetition. You should present varied opportunities to approach each mathematics concept from a variety of angles. You should also encourage reinforcement of each previously mastered concept along with the presentation of each new concept. This allows the child to build upon a solid foundation of comprehension that will encourage a true mastery of mathematics throughout their educational career.

The focus on series of lessons bring a laser focus to each subject area, ensuring each necessary skill is fully developed before the student moves onto the next skill in the series.

Continuity is critical to educational success. By ensuring each student gains a comprehensive understanding of each skill before transitioning to the next skill, you lay the groundwork for lifelong academic success.

UNIT 1 – MONTHS

A calendar is a way to keep track of the days of the week and the months of the year. Calendars help us to keep a schedule, know when special days occur, and plan activities.

Each year, we start a completely new calendar. There will be 12 pages to the calendar for the year. Each of these pages will show a specific month.

The months of the year are

January, February, March and April

May and June and July and August

September, October, November December

That's our year!

Keep singing that song until you remember every month of the year. Knowing the months of the year and the order of the months helps you to know when important events are happening.

Cut out the flashcards. Use them to help you memorize the months.

January	February	March
April	May	June
July	August	September
October	November	December

Name: ____

Cut out the mixed up months and then glue the months of the year back into the right order.

1	2	3
4	5	6
7	8	9
10	11	12

March	December	April
November	January	August
July	February	October
May	September	June

Name: ______________________________

Fill in the missing months.

1. January, February, ______________

2. October, ______________, December

3. ______________, March, April

4. June, ______________, August

5. March, April, ______________

6. ______________, October, November

7. April, May, ______________

8. ______________, February, March

9. July, ______________, September

10. October, November, ______________

11. September, ______________, November

12. March, ______________, May

Name: ______________________________

Use the calendar to answer the questions.

January						
Su	M	T	W	Th	F	Sa
		1	2	3	4	5
6	7	8	9	10	11	12
13	14	15	16	17	18	19
20	21	22	23	24	25	26
27	28	29	30	31		

February						
Su	M	T	W	Th	F	Sa
					1	2
3	4	5	6	7	8	9
10	11	12	13	14	15	16
17	18	19	20	21	22	23
24	25	26	27	28		

March						
Su	M	T	W	Th	F	Sa
					1	2
3	4	5	6	7	8	9
10	11	12	13	14	15	16
17	18	19	20	21	22	23
24	25	26	27	28	29	30
31						

April						
Su	M	T	W	Th	F	Sa
	1	2	3	4	5	6
7	8	9	10	11	12	13
14	15	16	17	18	19	20
21	22	23	24	25	26	27
28	29	30				

May						
Su	M	T	W	Th	F	Sa
			1	2	3	4
5	6	7	8	9	10	11
12	13	14	15	16	17	18
19	20	21	22	23	24	25
26	27	28	29	30	31	

June						
Su	M	T	W	Th	F	Sa
						1
2	3	4	5	6	7	8
9	10	11	12	13	14	15
16	17	18	19	20	21	22
23	24	25	26	27	28	29
30						

July						
Su	M	T	W	Th	F	Sa
	1	2	3	4	5	6
7	8	9	10	11	12	13
14	15	16	17	18	19	20
21	22	23	24	25	26	27
28	29	30	31			

August						
Su	M	T	W	Th	F	Sa
				1	2	3
4	5	6	7	8	9	10
11	12	13	14	15	16	17
18	19	20	21	22	23	24
25	26	27	28	29	30	31

September						
Su	M	T	W	Th	F	Sa
1	2	3	4	5	6	7
8	9	10	11	12	13	14
15	16	17	18	19	20	21
22	23	24	25	26	27	28
29	30					

October						
Su	M	T	W	Th	F	Sa
		1	2	3	4	5
6	7	8	9	10	11	12
13	14	15	16	17	18	19
20	21	22	23	24	25	26
27	28	29	30	31		

November						
Su	M	T	W	Th	F	Sa
					1	2
3	4	5	6	7	8	9
10	11	12	13	14	15	16
17	18	19	20	21	22	23
24	25	26	27	28	29	30

December						
Su	M	T	W	Th	F	Sa
1	2	3	4	5	6	7
8	9	10	11	12	13	14
15	16	17	18	19	20	21
22	23	24	25	26	27	28
29	30	31				

1. How many months are in a year? ______________________

2. What month comes before November? ______________________

3. What month comes after February? ______________________

4. What month comes between April and June? ______________________

5. What is the first month of the year? ______________________

Name: ____

Use the calendar to answer the questions.

January

Su	M	T	W	Th	F	Sa
		1	2	3	4	5
6	7	8	9	10	11	12
13	14	15	16	17	18	19
20	21	22	23	24	25	26
27	28	29	30	31		

February

Su	M	T	W	Th	F	Sa
					1	2
3	4	5	6	7	8	9
10	11	12	13	14	15	16
17	18	19	20	21	22	23
24	25	26	27	28		

March

Su	M	T	W	Th	F	Sa
					1	2
3	4	5	6	7	8	9
10	11	12	13	14	15	16
17	18	19	20	21	22	23
24	25	26	27	28	29	30
31						

April

Su	M	T	W	Th	F	Sa
	1	2	3	4	5	6
7	8	9	10	11	12	13
14	15	16	17	18	19	20
21	22	23	24	25	26	27
28	29	30				

May

Su	M	T	W	Th	F	Sa
			1	2	3	4
5	6	7	8	9	10	11
12	13	14	15	16	17	18
19	20	21	22	23	24	25
26	27	28	29	30	31	

June

Su	M	T	W	Th	F	Sa
						1
2	3	4	5	6	7	8
9	10	11	12	13	14	15
16	17	18	19	20	21	22
23	24	25	26	27	28	29
30						

July

Su	M	T	W	Th	F	Sa
	1	2	3	4	5	6
7	8	9	10	11	12	13
14	15	16	17	18	19	20
21	22	23	24	25	26	27
28	29	30	31			

August

Su	M	T	W	Th	F	Sa
				1	2	3
4	5	6	7	8	9	10
11	12	13	14	15	16	17
18	19	20	21	22	23	24
25	26	27	28	29	30	31

September

Su	M	T	W	Th	F	Sa
1	2	3	4	5	6	7
8	9	10	11	12	13	14
15	16	17	18	19	20	21
22	23	24	25	26	27	28
29	30					

October

Su	M	T	W	Th	F	Sa
		1	2	3	4	5
6	7	8	9	10	11	12
13	14	15	16	17	18	19
20	21	22	23	24	25	26
27	28	29	30	31		

November

Su	M	T	W	Th	F	Sa
					1	2
3	4	5	6	7	8	9
10	11	12	13	14	15	16
17	18	19	20	21	22	23
24	25	26	27	28	29	30

December

Su	M	T	W	Th	F	Sa
1	2	3	4	5	6	7
8	9	10	11	12	13	14
15	16	17	18	19	20	21
22	23	24	25	26	27	28
29	30	31				

1. What is the last month of the year? ______________________________
2. What month comes before April? ______________________________
3. What month comes after August? ______________________________
4. What month comes between May and July? ______________________
5. What month comes after January? ______________________________

Name: ____

Use the calendar to complete each sentence.

January

Su	M	T	W	Th	F	Sa
		1	2	3	4	5
6	7	8	9	10	11	12
13	14	15	16	17	18	19
20	21	22	23	24	25	26
27	28	29	30	31		

February

Su	M	T	W	Th	F	Sa
					1	2
3	4	5	6	7	8	9
10	11	12	13	14	15	16
17	18	19	20	21	22	23
24	25	26	27	28		

March

Su	M	T	W	Th	F	Sa
					1	2
3	4	5	6	7	8	9
10	11	12	13	14	15	16
17	18	19	20	21	22	23
24	25	26	27	28	29	30
31						

April

Su	M	T	W	Th	F	Sa
	1	2	3	4	5	6
7	8	9	10	11	12	13
14	15	16	17	18	19	20
21	22	23	24	25	26	27
28	29	30				

May

Su	M	T	W	Th	F	Sa
			1	2	3	4
5	6	7	8	9	10	11
12	13	14	15	16	17	18
19	20	21	22	23	24	25
26	27	28	29	30	31	

June

Su	M	T	W	Th	F	Sa
						1
2	3	4	5	6	7	8
9	10	11	12	13	14	15
16	17	18	19	20	21	22
23	24	25	26	27	28	29
30						

July

Su	M	T	W	Th	F	Sa
	1	2	3	4	5	6
7	8	9	10	11	12	13
14	15	16	17	18	19	20
21	22	23	24	25	26	27
28	29	30	31			

August

Su	M	T	W	Th	F	Sa
				1	2	3
4	5	6	7	8	9	10
11	12	13	14	15	16	17
18	19	20	21	22	23	24
25	26	27	28	29	30	31

September

Su	M	T	W	Th	F	Sa
1	2	3	4	5	6	7
8	9	10	11	12	13	14
15	16	17	18	19	20	21
22	23	24	25	26	27	28
29	30					

October

Su	M	T	W	Th	F	Sa
		1	2	3	4	5
6	7	8	9	10	11	12
13	14	15	16	17	18	19
20	21	22	23	24	25	26
27	28	29	30	31		

November

Su	M	T	W	Th	F	Sa
					1	2
3	4	5	6	7	8	9
10	11	12	13	14	15	16
17	18	19	20	21	22	23
24	25	26	27	28	29	30

December

Su	M	T	W	Th	F	Sa
1	2	3	4	5	6	7
8	9	10	11	12	13	14
15	16	17	18	19	20	21
22	23	24	25	26	27	28
29	30	31				

1. September is the ____________________ month of the year.

2. June is the ____________________ month of the year.

3. April is the ____________________ month of the year.

4. December is the ____________________ month of the year.

5. August is the ____________________ month of the year.

Name: ____

Use the calendar to complete each sentence.

January

Su	M	T	W	Th	F	Sa
		1	2	3	4	5
6	7	8	9	10	11	12
13	14	15	16	17	18	19
20	21	22	23	24	25	26
27	28	29	30	31		

February

Su	M	T	W	Th	F	Sa
					1	2
3	4	5	6	7	8	9
10	11	12	13	14	15	16
17	18	19	20	21	22	23
24	25	26	27	28		

March

Su	M	T	W	Th	F	Sa
					1	2
3	4	5	6	7	8	9
10	11	12	13	14	15	16
17	18	19	20	21	22	23
24	25	26	27	28	29	30
31						

April

Su	M	T	W	Th	F	Sa
	1	2	3	4	5	6
7	8	9	10	11	12	13
14	15	16	17	18	19	20
21	22	23	24	25	26	27
28	29	30				

May

Su	M	T	W	Th	F	Sa
			1	2	3	4
5	6	7	8	9	10	11
12	13	14	15	16	17	18
19	20	21	22	23	24	25
26	27	28	29	30	31	

June

Su	M	T	W	Th	F	Sa
						1
2	3	4	5	6	7	8
9	10	11	12	13	14	15
16	17	18	19	20	21	22
23	24	25	26	27	28	29
30						

July

Su	M	T	W	Th	F	Sa
	1	2	3	4	5	6
7	8	9	10	11	12	13
14	15	16	17	18	19	20
21	22	23	24	25	26	27
28	29	30	31			

August

Su	M	T	W	Th	F	Sa
				1	2	3
4	5	6	7	8	9	10
11	12	13	14	15	16	17
18	19	20	21	22	23	24
25	26	27	28	29	30	31

September

Su	M	T	W	Th	F	Sa
1	2	3	4	5	6	7
8	9	10	11	12	13	14
15	16	17	18	19	20	21
22	23	24	25	26	27	28
29	30					

October

Su	M	T	W	Th	F	Sa
		1	2	3	4	5
6	7	8	9	10	11	12
13	14	15	16	17	18	19
20	21	22	23	24	25	26
27	28	29	30	31		

November

Su	M	T	W	Th	F	Sa
					1	2
3	4	5	6	7	8	9
10	11	12	13	14	15	16
17	18	19	20	21	22	23
24	25	26	27	28	29	30

December

Su	M	T	W	Th	F	Sa
1	2	3	4	5	6	7
8	9	10	11	12	13	14
15	16	17	18	19	20	21
22	23	24	25	26	27	28
29	30	31				

1. March is the ______________________ month of the year.

2. February is the ____________________ month of the year.

3. July is the ________________________ month of the year.

4. May is the ________________________ month of the year.

5. January is the _____________________ month of the year.

UNIT 2 – DAYS OF THE WEEK

Each month is divided into weeks. There are 52 weeks in a year. Each week has 7 days.

Sunday, Monday, Tuesday, Wednesday

Thursday, Friday, Saturday and back to Sunday

Cut out the flashcards. Use them to help you memorize the days of the week.

Sunday
Monday
Tuesday
Wednesday
Thursday
Friday
Saturday

Name: ______________________________

Cut out the mixed up days of the week and then glue the days of the week back into the right order.

Thursday	1
Monday	2
Saturday	3
Friday	4
Tuesday	5
Sunday	6
Wednesday	7

Name: ____

Fill in the missing day.

1. Sunday, __________________, Tuesday

2. Thursday, Friday, __________________

3. __________________, Monday, Tuesday

4. Tuesday, __________________, Thursday

5. __________________, Wednesday, Thursday

6. Thursday, __________________, Saturday

7. Tuesday, Wednesday, __________________

8. __________________, Tuesday, Wednesday

9. Wednesday, __________________, Friday

10. Monday, Tuesday, __________________

Name: ____

Use your ordered flashcards to answer the questions.

1. How many days are in a week? ____________________

2. What is the first day of the week? ____________________

3. What day comes after Friday? ____________________

4. What day comes after Tuesday? ____________________

5. What day comes before Wednesday? ____________________

6. What day comes before Monday? ____________________

7. What day comes before Saturday? ____________________

8. What day comes after Monday? ____________________

9. What day comes after Thursday? ____________________

10. What day comes before Friday? ____________________

11. What day comes before Tuesday? ____________________

12. What is the last day of the week? ____________________

13. What day comes before Thursday? ____________________

14. What day comes after Sunday? ____________________

Name: ______________________________

Use your ordered flashcards to complete each sentence.
Answer using number words. Hint: First

1. Wednesday is the ____________________ day of the week.
2. Saturday is the ____________________ day of the week.
3. Monday is the ____________________ day of the week.
4. Tuesday is the ____________________ day of the week.
5. Thursday is the ____________________ day of the week.
6. Sunday is the ____________________ day of the week.
7. Friday is the ____________________ day of the week.

UNIT 3 – PARTS OF A CALENDAR

Month

Each page of the calendar will show a month. Each month gets its own page. The top line of the calendar page will show the name of the month.

January						
Sunday	Monday	Tuesday	Wednesday	Thursday	Friday	Saturday
1	2	3	4	5	6	7

January, February, March, and April.

May and June and July and August.

September, October, November, December.

That's our year!

Days of the Week

The second row of the calendar will show the days of the week.

The calendar starts on Sunday and ends on Saturday.

January						
Sunday	**Monday**	**Tuesday**	**Wednesday**	**Thursday**	**Friday**	**Saturday**
1	2	3	4	5	6	7

Date

The next rows of the calendar will have numbers. These numbers line up with the day columns. Each day of the month has a number. These numbers are the dates.

January						
Sunday	Monday	Tuesday	Wednesday	Thursday	Friday	Saturday
1	**2**	**3**	**4**	**5**	**6**	**7**

Each month will start and end on a different day of the week. The months have different numbers of days. A simple rhyme can help you remember how many days are in each month.

Thirty days has September,

April, June, and November.

All the rest have 31,

except February alone,

which has only 28 days clear

and 29 in each leap year.

Name: ____

Fill in the blanks to the poem.

Thirty days has ________________,

________________, June, and ________________.

All the rest have ________________,

except ___________________alone,

which has only ________________ days clear

and ___________________ in each leap year.

Name: ____

Write how many days each month has on the line.

1.	January	______	7.	July	______
2.	February	______	8.	August	______
3.	March	______	9.	September	______
4.	April	______	10.	October	______
5.	May	______	11.	November	______
6.	June	______	12.	December	______

Name: ____

Use the calendar to complete each sentence.

January

Su	M	T	W	Th	F	Sa
		1	2	3	4	5
6	7	8	9	10	11	12
13	14	15	16	17	18	19
20	21	22	23	24	25	26
27	28	29	30	31		

February

Su	M	T	W	Th	F	Sa
					1	2
3	4	5	6	7	8	9
10	11	12	13	14	15	16
17	18	19	20	21	22	23
24	25	26	27	28		

March

Su	M	T	W	Th	F	Sa
					1	2
3	4	5	6	7	8	9
10	11	12	13	14	15	16
17	18	19	20	21	22	23
24	25	26	27	28	29	30
31						

April

Su	M	T	W	Th	F	Sa
	1	2	3	4	5	6
7	8	9	10	11	12	13
14	15	16	17	18	19	20
21	22	23	24	25	26	27
28	29	30				

May

Su	M	T	W	Th	F	Sa
			1	2	3	4
5	6	7	8	9	10	11
12	13	14	15	16	17	18
19	20	21	22	23	24	25
26	27	28	29	30	31	

June

Su	M	T	W	Th	F	Sa
						1
2	3	4	5	6	7	8
9	10	11	12	13	14	15
16	17	18	19	20	21	22
23	24	25	26	27	28	29
30						

July

Su	M	T	W	Th	F	Sa
	1	2	3	4	5	6
7	8	9	10	11	12	13
14	15	16	17	18	19	20
21	22	23	24	25	26	27
28	29	30	31			

August

Su	M	T	W	Th	F	Sa
				1	2	3
4	5	6	7	8	9	10
11	12	13	14	15	16	17
18	19	20	21	22	23	24
25	26	27	28	29	30	31

September

Su	M	T	W	Th	F	Sa
1	2	3	4	5	6	7
8	9	10	11	12	13	14
15	16	17	18	19	20	21
22	23	24	25	26	27	28
29	30					

October

Su	M	T	W	Th	F	Sa
		1	2	3	4	5
6	7	8	9	10	11	12
13	14	15	16	17	18	19
20	21	22	23	24	25	26
27	28	29	30	31		

November

Su	M	T	W	Th	F	Sa
					1	2
3	4	5	6	7	8	9
10	11	12	13	14	15	16
17	18	19	20	21	22	23
24	25	26	27	28	29	30

December

Su	M	T	W	Th	F	Sa
1	2	3	4	5	6	7
8	9	10	11	12	13	14
15	16	17	18	19	20	21
22	23	24	25	26	27	28
29	30	31				

1. How many months have 31 days? ______________________

2. How many days are in December? ______________________

3. How many days are in August? ______________________

4. How many days are in July? ______________________

5. How many days are in January? ______________________

Name: ____

Use the calendar to complete each sentence.

January						
Su	M	T	W	Th	F	Sa
		1	2	3	4	5
6	7	8	9	10	11	12
13	14	15	16	17	18	19
20	21	22	23	24	25	26
27	28	29	30	31		

February						
Su	M	T	W	Th	F	Sa
					1	2
3	4	5	6	7	8	9
10	11	12	13	14	15	16
17	18	19	20	21	22	23
24	25	26	27	28		

March						
Su	M	T	W	Th	F	Sa
					1	2
3	4	5	6	7	8	9
10	11	12	13	14	15	16
17	18	19	20	21	22	23
24	25	26	27	28	29	30
31						

April						
Su	M	T	W	Th	F	Sa
	1	2	3	4	5	6
7	8	9	10	11	12	13
14	15	16	17	18	19	20
21	22	23	24	25	26	27
28	29	30				

May						
Su	M	T	W	Th	F	Sa
			1	2	3	4
5	6	7	8	9	10	11
12	13	14	15	16	17	18
19	20	21	22	23	24	25
26	27	28	29	30	31	

June						
Su	M	T	W	Th	F	Sa
						1
2	3	4	5	6	7	8
9	10	11	12	13	14	15
16	17	18	19	20	21	22
23	24	25	26	27	28	29
30						

July						
Su	M	T	W	Th	F	Sa
	1	2	3	4	5	6
7	8	9	10	11	12	13
14	15	16	17	18	19	20
21	22	23	24	25	26	27
28	29	30	31			

August						
Su	M	T	W	Th	F	Sa
				1	2	3
4	5	6	7	8	9	10
11	12	13	14	15	16	17
18	19	20	21	22	23	24
25	26	27	28	29	30	31

September						
Su	M	T	W	Th	F	Sa
1	2	3	4	5	6	7
8	9	10	11	12	13	14
15	16	17	18	19	20	21
22	23	24	25	26	27	28
29	30					

October						
Su	M	T	W	Th	F	Sa
		1	2	3	4	5
6	7	8	9	10	11	12
13	14	15	16	17	18	19
20	21	22	23	24	25	26
27	28	29	30	31		

November						
Su	M	T	W	Th	F	Sa
					1	2
3	4	5	6	7	8	9
10	11	12	13	14	15	16
17	18	19	20	21	22	23
24	25	26	27	28	29	30

December						
Su	M	T	W	Th	F	Sa
1	2	3	4	5	6	7
8	9	10	11	12	13	14
15	16	17	18	19	20	21
22	23	24	25	26	27	28
29	30	31				

1. How many days are in October? ______________________________

2. How many days are in February? ______________________________

3. How many months have 30 days? ______________________________

4. How many days are in November? ______________________________

5. How many days are in June? ______________________________

Name: ______________________________

Use the calendar to complete each sentence.

January

Su	M	T	W	Th	F	Sa
		1	2	3	4	5
6	7	8	9	10	11	12
13	14	15	16	17	18	19
20	21	22	23	24	25	26
27	28	29	30	31		

February

Su	M	T	W	Th	F	Sa
					1	2
3	4	5	6	7	8	9
10	11	12	13	14	15	16
17	18	19	20	21	22	23
24	25	26	27	28		

March

Su	M	T	W	Th	F	Sa
					1	2
3	4	5	6	7	8	9
10	11	12	13	14	15	16
17	18	19	20	21	22	23
24	25	26	27	28	29	30
31						

April

Su	M	T	W	Th	F	Sa
	1	2	3	4	5	6
7	8	9	10	11	12	13
14	15	16	17	18	19	20
21	22	23	24	25	26	27
28	29	30				

May

Su	M	T	W	Th	F	Sa
			1	2	3	4
5	6	7	8	9	10	11
12	13	14	15	16	17	18
19	20	21	22	23	24	25
26	27	28	29	30	31	

June

Su	M	T	W	Th	F	Sa
						1
2	3	4	5	6	7	8
9	10	11	12	13	14	15
16	17	18	19	20	21	22
23	24	25	26	27	28	29
30						

July

Su	M	T	W	Th	F	Sa
	1	2	3	4	5	6
7	8	9	10	11	12	13
14	15	16	17	18	19	20
21	22	23	24	25	26	27
28	29	30	31			

August

Su	M	T	W	Th	F	Sa
				1	2	3
4	5	6	7	8	9	10
11	12	13	14	15	16	17
18	19	20	21	22	23	24
25	26	27	28	29	30	31

September

Su	M	T	W	Th	F	Sa
1	2	3	4	5	6	7
8	9	10	11	12	13	14
15	16	17	18	19	20	21
22	23	24	25	26	27	28
29	30					

October

Su	M	T	W	Th	F	Sa
		1	2	3	4	5
6	7	8	9	10	11	12
13	14	15	16	17	18	19
20	21	22	23	24	25	26
27	28	29	30	31		

November

Su	M	T	W	Th	F	Sa
					1	2
3	4	5	6	7	8	9
10	11	12	13	14	15	16
17	18	19	20	21	22	23
24	25	26	27	28	29	30

December

Su	M	T	W	Th	F	Sa
1	2	3	4	5	6	7
8	9	10	11	12	13	14
15	16	17	18	19	20	21
22	23	24	25	26	27	28
29	30	31				

1. How many days are in March? ______________________

2. How many days are in May? ______________________

3. How many months have 28 days? ______________________

4. How many days are in April? ______________________

5. How many days are in September? ______________________

Name: ______________________________

Use the calendar to answer the questions.

MAY						
Sunday	Monday	Tuesday	Wednesday	Thursday	Friday	Saturday
			1	2	3	4
5	6	7	8	9	10	11
12	13	14	15	16	17	18
19	20	21	22	23	24	25
26	27	28	29	30	31	

1. What day of the week is May 8th? ______________________

2. How many Wednesdays are in May? ______________________

3. What date is the 3rd Friday in May? ______________________

4. How many days are in May? ______________________

5. What is the first day in May? ______________________

6. How many Saturdays are in May? ______________________

7. What date is the first Monday in May? ______________________

8. What day of the week is May 23rd? ______________________

Name: ______________________________

Use the calendar to answer the questions.

October						
Sunday	Monday	Tuesday	Wednesday	Thursday	Friday	Saturday
		1	2	3	4	5
6	7	8	9	10	11	12
13	14	15	16	17	18	19
20	21	22	23	24	25	26
27	28	29	30	31		

1. What day of the week is October 16th? ______________________
2. How many Fridays are in October? ______________________
3. What date is the 2nd Tuesday in October? ______________________
4. How many days are in October? ______________________
5. What is the last day in October? ______________________
6. How many Mondays are in October? ______________________
7. What date is the 4th Monday in October? ______________________
8. What day of the week is October 9th? ______________________

Name: ______________________________

Use the calendar to answer the questions.

February						
Sunday	Monday	Tuesday	Wednesday	Thursday	Friday	Saturday
					1	2
3	4	5	6	7	8	9
10	11	12	13	14	15	16
17	18	19	20	21	22	23
24	25	26	27	28		

1. What day of the week is February 12th? ______________________
2. How many Mondays are in February? ______________________
3. What date is the 1st Wednesday in February? ______________________
4. How many days are in February? ______________________
5. What is the first day in February? ______________________
6. How many Fridays are in February? ______________________
7. What date is the 2nd Tuesday in February? ______________________
8. What day of the week is February 6th? ______________________

Name: ______________________________

Use the calendar to answer the questions.

November						
Sunday	Monday	Tuesday	Wednesday	Thursday	Friday	Saturday
					1	2
3	4	5	6	7	8	9
10	11	12	13	14	15	16
17	18	19	20	21	22	23
24	25	26	27	28	29	30

1. What day of the week is November 7th? ______________________
2. How many Sundays are in November? ______________________
3. What date is the 2nd Thursday in November? ______________________
4. How many days are in November? ______________________
5. What is the first day in November? ______________________
6. How many Tuesdays are in November? ______________________
7. What date is the 3rd Monday in November? ______________________
8. What day of the week is November 28th? ______________________

UNIT 4 – MAKE A CALENDAR

You can make a calendar that you can use this year!

Start by remembering the order of the months.

January, February, March, and April

May and June and July and August

September, October, November, December

That's our year!

The first month is January

1. Add the name of the month to the top of the first calendar page.

2. Number the days beginning with the 1st day of the month.

   Make sure that you put the 1st day of the month in the right square. This will be the top row of your chart but the day column you use will change each year.

   Find out what day of the week New Year's Day fell on this year.

   That will be the very first day of your month and the day that will help you to figure out the rest of your calendar.

   In 2015, New Year's Day was a Thursday and in 2016 New Year's Day was a Friday.

   You need to remember how many days are in each month so that you know when to stop writing numbers on your calendar.

30 days has September,

April, June and November

All the rest have 31

Except great February with 28

but Leap Year's February is fine with 29

What the song means is that April, June, September, and November have 30 days.

January, March, May, July, August, October, and December have 31 days.

February usually has 28 days but every 4 years we add a day to February. We call this a leap year because we leap an extra day to get through the year.

3. January has 31 days, so keep adding numbers to the boxes until you reach 31.

Sunday	Monday	Tuesday	Wednesday	Thursday	Friday	Saturday

Make a Calendar - February

The next month is February.

1. Add the name of the month to the top row of the calendar.

2. Start numbering your days with the 1st day of the month.

   Make sure that you put the 1st day of the month in the right square. This will be the top row of your chart but the day column you use will change each year.

   The first day of February will be the day after the last day of January.

   Look at your January calendar.

   What day was the last day of the month? ____________________

   The first day of February will be the next day. __________________

3. Keep numbering until you have made a box for each day.

   February is an unusual month. The number of days in February can change. February is the only month whose number of days change. This change happens because of something called a leap year.

   A leap year is a year that has one extra day in it. Most years have 365 days.

   We define our year by the amount of time it takes the Earth to orbit the sun. It actually takes 365 1/4 days for the earth to orbit the sun.

   We could not make a calendar that has 365 1/4 days in it so we save the extra 1/4 day until we have reached a full extra day. We then add the full extra day to the shortest month of the year, February. It takes four 1/4's to equal a whole, so we add a day once every four years making that year a leap year.

Sunday	Monday	Tuesday	Wednesday	Thursday	Friday	Saturday

Make a Calendar - March

The next month is March.

1. Add the name of the month to the top row of the calendar.

2. Start numbering your days with the 1st day of the month.

   Make sure that you put the 1st day of the month in the right square.

   This will be the top row of your chart but the day column you use will change each year.

   The first day of March will be the day after the last day of February.

   Look at your February calendar.

   What day was the last day of the month? ______________________

   The first day of March will be the next day. ______________________

3. Keep numbering until you have made a box for each day.

   How many days are in March? ___________

Sunday	Monday	Tuesday	Wednesday	Thursday	Friday	Saturday

Make a Calendar - April

The next month is April.

1. Add the name of the month to the top row of the calendar.

2. Start numbering your days with the 1st day of the month.

   Make sure that you put the 1st day of the month in the right square.

   This will be the top row of your chart but the day column you use will change each year.

   The first day of April will be the day after the last day of March.

   Look at your March calendar.

   What day was the last day of the month? ____________________

   The first day of April will be the next day. ____________________

3. Keep numbering until you have made a box for each day.

   How many days are in April? ____________

Sunday	Monday	Tuesday	Wednesday	Thursday	Friday	Saturday

Make a Calendar - May

The next month is May.

1. Add the name of the month to the top row of the calendar.

2. Start numbering your days with the 1st day of the month.

   Make sure that you put the 1st day of the month in the right square.

   This will be the top row of your chart but the day column you use will change each year.

   The first day of May will be the day after the last day of April.

   Look at your April calendar.

   What day was the last day of the month? ____________________

   The first day of May will be the next day. ____________________

3. Keep numbering until you have made a box for each day.

   How many days are in May? ____________

Sunday	Monday	Tuesday	Wednesday	Thursday	Friday	Saturday

Make a Calendar - June

The next month is June.

1. Add the name of the month to the top row of the calendar.

2. Start numbering your days with the 1st day of the month.

   Make sure that you put the 1st day of the month in the right square.

   This will be the top row of your chart but the day column you use will change each year.

   The first day of June will be the day after the last day of May.

   Look at your May calendar.

   What day was the last day of the month? ______________________________

   The first day of June will be the next day. ______________________________

3. Keep numbering until you have made a box for each day.

   How many days are in June? __________________

Sunday	Monday	Tuesday	Wednesday	Thursday	Friday	Saturday

Make a Calendar - July

The next month is July.

1. Add the name of the month to the top row of the calendar.

2. Start numbering your days with the 1st day of the month.

   Make sure that you put the 1st day of the month in the right square.

   This will be the top row of your chart but the day column you use will change each year.

   The first day of July will be the day after the last day of June.

   Look at your June calendar.

   What day was the last day of the month? ______________________

   The first day of July will be the next day. ______________________

3. Keep numbering until you have made a box for each day.

   How many days are in July? ______________

Sunday	Monday	Tuesday	Wednesday	Thursday	Friday	Saturday

Make a Calendar - August

The next month is August.

1. Add the name of the month to the top row of the calendar.

2. Start numbering your days with the 1st day of the month.

   Make sure that you put the 1st day of the month in the right square.

   This will be the top row of your chart but the day column you use will change each year.

   The first day of August will be the day after the last day of July.

   Look at your July calendar.

   What day was the last day of the month? ____________________

   The first day of August will be the next day. ____________________

3. Keep numbering until you have made a box for each day.

   How many days are in August? ____________

Sunday	Monday	Tuesday	Wednesday	Thursday	Friday	Saturday

Make a Calendar - September

The next month is September.

1. Add the name of the month to the top row of the calendar.

2. Start numbering your days with the 1st day of the month.

   Make sure that you put the 1st day of the month in the right square.

   This will be the top row of your chart but the day column you use will change each year.

   The first day of September will be the day after the last day of August.

   Look at your August calendar.

   What day was the last day of the month? ____________________

   The first day of September will be the next day. ________________

3. Keep numbering until you have made a box for each day.

   How many days are in September? ____________

Sunday	Monday	Tuesday	Wednesday	Thursday	Friday	Saturday

Make a Calendar - October

The next month is October.

1. Add the name of the month to the top row of the calendar.

2. Start numbering your days with the 1st day of the month.

   Make sure that you put the 1st day of the month in the right square.

   This will be the top row of your chart but the day column you use will change each year.

   The first day of October will be the day after the last day of September.

   Look at your September calendar.

   What day was the last day of the month? ______________________

   The first day of October will be the next day. ______________________

3. Keep numbering until you have made a box for each day.

   How many days are in October? ______________

Sunday	Monday	Tuesday	Wednesday	Thursday	Friday	Saturday

Make a Calendar - November

The next month is November.

1. Add the name of the month to the top row of the calendar.

2. Start numbering your days with the 1st day of the month.

   Make sure that you put the 1st day of the month in the right square.

   This will be the top row of your chart but the day column you use will change each year.

   The first day of November will be the day after the last day of October.

   Look at your October calendar.

   What day was the last day of the month? ____________________

   The first day of November will be the next day. ____________________

3. Keep numbering until you have made a box for each day.

   How many days are in November? ____________

Sunday	Monday	Tuesday	Wednesday	Thursday	Friday	Saturday

Make a Calendar - December

The last month of the year is December.

1. Add the name of the month to the top row of the calendar.

2. Start numbering your days with the 1st day of the month.

   Make sure that you put the 1st day of the month in the right square.

   This will be the top row of your chart but the day column you use will change each year.

   The first day of December will be the day after the last day of November.

   Look at your November calendar.

   What day was the last day of the month? ____________________

   The first day of December will be the next day. ________________

3. Keep numbering until you have made a box for each day.

   How many days are in December? ____________

Sunday	Monday	Tuesday	Wednesday	Thursday	Friday	Saturday

UNIT 5 – SPECIAL DAYS

Some dates will be marked with a note. These notes tell you about special events or holidays that happens on that day.

January						
Sunday	Monday	Tuesday	Wednesday	Thursday	Friday	Saturday
1 New Year's Day	2	3	4	5	6	7

You should have a calendar page for each month of the year.

Cut out each calendar page and put them in order. Staple the top of your new calendar so that you can use it throughout the year to keep track of holidays, special days, appointments and other important events.

It is time to add holidays and other special days to your calendar.

Holidays are funny. Some of them happen on exactly the same date each year. Others change dates but happen on the same day of the week each year.

January has 2 holidays. You should look to see what the date of each holiday is for the year you are making your calendar.

In 2016, the holidays for January might have different dates than in 2018 or 2021. Some holidays are on very specific dates. Other holidays fall on a specific day of the week. Since the day of the week for each date changes depending on the first day of the year, these holidays will have different dates!

New Year's Day is always on January 1st. Christmas is always December 25th. Other holidays move around depending on the calendar. Memorial Day is observed on the last Monday in May. Since the days of the month move each year, the last Monday in May can be a different date this year than it will be next year.

You should look at the dates for each holiday before you add them to your calendar.

Common Holiday List

New Year's Day
New Year's Day is always celebrated on January 1st.

Martin Luther King Jr. Birthday
Martin Luther King Junior's actual birthday is January 15th but we celebrate it on the Monday after his birthday. The date will move each year depending on when the next Monday happens.

Groundhog Day
Groundhog's Day is always celebrated on February 2nd.

Valentine's Day
Valentine's Day is always celebrated on February $14^{th.}$

President's Day
We celebrate the Birthday's of some of the Nations most remembered Presidents. Instead of celebrating each Birthday separately, we made President's Day the 3rd Monday in February.

Daylight Saving Time begins
Daylight Savings Time starts on the 2nd Sunday in March at 2 a.m. The date will change each year.

Daylight savings time is when our county moves clocks forward by one hour during the spring as the days grow longer. Adjusting the clocks makes the evening last longer.

We adjust the clocks backward one hour in the fall when the days grow shorter. This practice makes it feel as if it is getting light earlier so that people who have to go to work or catch a bus to school early in the morning are not doing it in the dark.

A good way to remember daylight savings time is "spring" ahead and "fall" back.

St. Patrick's Day
St. Patrick's Day is always celebrated on March 17th.

April Fools Day
April Fools Day is always celebrated on April 1st.

Easter
Easter is always celebrated on a Sunday on or after March 22nd. Easter follows rules known as the Ecclesiastical Rules and will have a different date each year. You will need to ask what date Easter Sunday is for your year before completing your calendar.

Mothers Day
Mother's Day is always celebrated on the 2nd Sunday in May.

Memorial Day
Memorial Day is always celebrated on the last Monday in May.

Father's Day
Father's Day is always celebrated on the 3rd Sunday in June.

Independence Day
Independence Day is always celebrated on the 4th of July.

Labor Day
Labor Day is always celebrated on the 1st Monday in September.

Columbus Day
Columbus Day is always celebrated on the 2nd Monday in October.

Halloween
Halloween is always celebrated on October 31st.

Daylight Saving Time Ends
Daylight Savings Time ends on the 1st Sunday in November at 2 a.m. The date will change each year.

Daylight savings time is when our county moves clocks forward by one hour during the spring as the days grow longer. Adjusting the clocks makes the evening last longer.

We adjust the clocks backward one hour in the fall when the days grow shorter. This practice makes it feel as if it is getting light earlier so that people who have to go to work or catch a bus to school early in the morning are not doing it in the dark.

A good way to remember daylight savings time is "spring" ahead and "fall" back.

Election Day

Election day is always on the 1st Tuesday on or after November 2nd.

Veteran's Day

Veteran's Day is always celebrated on November 11th. Sometimes we get an extra day off for Veteran's Day. If November 11th is a Saturday, we celebrate it on the Friday before. If November 11th is a Sunday, we celebrate it on the Monday after. This gives everyone a day off to remember those who have fought for our country.

Thanksgiving Day

Thanksgiving Day is always celebrated on the 4th Thursday in November.

Christmas

Christmas Day is always December 25th.

New Year's Eve

New Year's Eve is the last day of the year. The last day of any year will be December 31st.

You might have other special days that you want to put on your calendar like birthdays, school vacations, family vacations, or other events that are important to you and your family. Spend some time finding out what days are important just for your family and put them on the right date in the right month of your calendar.

Some people like to decorate the calendar page for each month to help them remember what that month will bring. Why don't you decorate your calendar so that the pictures for each month make you think of special things that happen during each month? You might draw fireworks for the 4th of July or butterflies for a summer month. Be creative. This is your calendar to use all year long.

UNIT 6 - WRITING DATES

Sometimes the date is written day of the week, month, date, and the year.

Tuesday, January 1, 2016 is New Year's Day.

Wednesday, December 25, 2016 is Christmas Day.

You will use dates to help keep your work organized, to help plan your yearly schedule, and to keep track of special days in your life.

You try it. Use your calendar to write the date for each special day.

Groundhog Day ____

Valentine's Day ____

President's Day ____

Mother's Day ____

Father's Day ____

Independence Day ____

Labor Day ____

Thanksgiving ____

Christmas ____

Your Birthday ____

Ways to Write Dates

Sometimes the date is written day of the week, month, date, and the year.

Tuesday, January 1, 2016

Sometimes, we leave off the day of the week and only write the month, date, and year.

January 1, 2016

Sometimes, we abbreviate the date and write it in numerical form. Do you remember when you pasted the months of the year to the number grid? These numbers are another way of referring to the months of the year.

Each month of the year matches a number.

01	January
02	February
03	March
04	April
05	May
06	June
07	July
08	August

09	September
10	October
11	November
12	December

We can use the number in place of the name of the month to write the date. When you use numbers for a date, you must separate them so the reader is not confused. You can separate the numbers with a / or a – symbol. Using a number January 1 would be written as 01-02.

01/01/2016

01-01-2016

Name: ______________________________

Write the date in number form.

1. January 10, 2012 ______________________________

2. November 4, 2021 ______________________________

3. December 29, 1999 ______________________________

4. June 24, 2071 ______________________________

5. April 13, 2020 ______________________________

6. October 30, 2011 ______________________________

7. February 19, 2002 ______________________________

8. August 23, 2006 ______________________________

9. March 3, 2033 ______________________________

10. May 11, 1902 ______________________________

11. July 4, 2017 ______________________________

12. September 28, 1954 ______________________________

Name: ______________________________

Write the date in long form.

1. 11-14-2017 ______________________________

2. 03-27-2006 ______________________________

3. 12-23-2021 ______________________________

4. 01-04-1999 ______________________________

5. 02-19-2002 ______________________________

6. 04-13-1970 ______________________________

7. 08-23-2006 ______________________________

8. 05-31-2071 ______________________________

9. 07-11-1965 ______________________________

10. 06-03-2070 ______________________________

11. 09-23-1952 ______________________________

12. 10-12-2015 ______________________________

UNIT 7 - USING ABBREVIATIONS

Sometimes, we abbreviate the names of the months of the year or the days of the week. These abbreviations make it easier and faster when we write dates. The abbreviations of the months of the year and days of the week use the first few letters of the full name unless it is a very short name. We do not abbreviate very short names.

MONTHS OF THE YEAR

January = Jan.

February = Feb.

March = Mar.

April = Apr.

May = May

June = June

July = July

August = Aug.

September = Sept.

October = Oct.

November = Nov.

December = Dec.

Name: ______________________________

You try it. Draw a line to match the month name with its abbreviation.

January	Mar.
February	July
March	Feb.
April	Dec.
May	Sept.
June	Nov.
July	May
August	Oct.
September	Aug.
October	June
November	Jan.
December	Apr.

Name: ______________________________

Test Yourself! See if you can write the abbreviation to go with each month.

January ______________________

February ______________________

March ______________________

April ______________________

May ______________________

June ______________________

July ______________________

August ______________________

September ______________________

October ______________________

November ______________________

December ______________________

Name: ______________________________

Using the abbreviation for the month, January 1st would be written as Jan. 01.

You try it, Use your calendar to write the date for each special day using the abbreviation for the month.

1. New Year's Day ______________________________

2. Daylight Savings Time Begins ______________________________

3. St. Patrick's Day ______________________________

4. Easter ______________________________

5. Memorial Day ______________________________

6. Halloween ______________________________

7. Veteran's Day ______________________________

8. New Year's Eve ______________________________

9. A special person's birthday ______________________________

10. A special person's birthday ______________________________

Name: ______________________________

Using the abbreviation for the month, January 1st would be written as Jan. 01.

You try it, Use your calendar to write the date for each Holiday using the number form for the month.

1. Martin Luther King Jr. Birthday ______________________

2. April Fools Day ______________________

3. Columbus Day ______________________

4. Daylight Savings Time Ends ______________________

5. Election Day ______________________

6. Christmas Eve ______________________

7. Labor Day ______________________

8. A special family day ______________________

9. Your parent's birthday ______________________

10. Your Birthday ______________________

ABBREVIATIONS FOR DAYS OF THE WEEK

Days of the week can also be abbreviated using the first few letters for each day's name.

Sunday	Sun.
Monday	Mon.
Tuesday	Tues.
Wednesday	Wed.
Thursday	Thurs.
Friday	Fri.
Saturday	Sat.

Name: ____________________

You try it. Write the abbreviation for each day of the week.

Sunday ____________________

Monday ____________________

Tuesday ____________________

Wednesday ____________________

Thursday ____________________

Friday ____________________

Saturday ____________________

Name: ______________________________

You try it. Write the day of the week for each abbreviation.

Sun. ______________________________

Mon. ______________________________

Tues. ______________________________

Wed. ______________________________

Thurs. ______________________________

Fri. ______________________________

Sat. ______________________________

Name: ______________________________

Use the calendar to answer the questions.

January

Su	M	T	W	Th	F	Sa
		1	2	3	4	5
6	7	8	9	10	11	12
13	14	15	16	17	18	19
20	21	22	23	24	25	26
27	28	29	30	31		

February

Su	M	T	W	Th	F	Sa
					1	2
3	4	5	6	7	8	9
10	11	12	13	14	15	16
17	18	19	20	21	22	23
24	25	26	27	28		

March

Su	M	T	W	Th	F	Sa
					1	2
3	4	5	6	7	8	9
10	11	12	13	14	15	16
17	18	19	20	21	22	23
24	25	26	27	28	29	30
31						

April

Su	M	T	W	Th	F	Sa
	1	2	3	4	5	6
7	8	9	10	11	12	13
14	15	16	17	18	19	20
21	22	23	24	25	26	27
28	29	30				

May

Su	M	T	W	Th	F	Sa
			1	2	3	4
5	6	7	8	9	10	11
12	13	14	15	16	17	18
19	20	21	22	23	24	25
26	27	28	29	30	31	

June

Su	M	T	W	Th	F	Sa
						1
2	3	4	5	6	7	8
9	10	11	12	13	14	15
16	17	18	19	20	21	22
23	24	25	26	27	28	29
30						

July

Su	M	T	W	Th	F	Sa
	1	2	3	4	5	6
7	8	9	10	11	12	13
14	15	16	17	18	19	20
21	22	23	24	25	26	27
28	29	30	31			

August

Su	M	T	W	Th	F	Sa
				1	2	3
4	5	6	7	8	9	10
11	12	13	14	15	16	17
18	19	20	21	22	23	24
25	26	27	28	29	30	31

September

Su	M	T	W	Th	F	Sa
1	2	3	4	5	6	7
8	9	10	11	12	13	14
15	16	17	18	19	20	21
22	23	24	25	26	27	28
29	30					

October

Su	M	T	W	Th	F	Sa
		1	2	3	4	5
6	7	8	9	10	11	12
13	14	15	16	17	18	19
20	21	22	23	24	25	26
27	28	29	30	31		

November

Su	M	T	W	Th	F	Sa
					1	2
3	4	5	6	7	8	9
10	11	12	13	14	15	16
17	18	19	20	21	22	23
24	25	26	27	28	29	30

December

Su	M	T	W	Th	F	Sa
1	2	3	4	5	6	7
8	9	10	11	12	13	14
15	16	17	18	19	20	21
22	23	24	25	26	27	28
29	30	31				

1. How many days are in a week? ______________

2. What month comes before May? ______________

3. What is the first holiday of the year? ______________

4. What month comes before August? ______________

5. What day comes after Monday? ______________

Name: ______________________________

Use the calendar to answer the questions.

January

Su	M	T	W	Th	F	Sa
		1	2	3	4	5
6	7	8	9	10	11	12
13	14	15	16	17	18	19
20	21	22	23	24	25	26
27	28	29	30	31		

February

Su	M	T	W	Th	F	Sa
					1	2
3	4	5	6	7	8	9
10	11	12	13	14	15	16
17	18	19	20	21	22	23
24	25	26	27	28		

March

Su	M	T	W	Th	F	Sa
					1	2
3	4	5	6	7	8	9
10	11	12	13	14	15	16
17	18	19	20	21	22	23
24	25	26	27	28	29	30
31						

April

Su	M	T	W	Th	F	Sa
	1	2	3	4	5	6
7	8	9	10	11	12	13
14	15	16	17	18	19	20
21	22	23	24	25	26	27
28	29	30				

May

Su	M	T	W	Th	F	Sa
			1	2	3	4
5	6	7	8	9	10	11
12	13	14	15	16	17	18
19	20	21	22	23	24	25
26	27	28	29	30	31	

June

Su	M	T	W	Th	F	Sa
						1
2	3	4	5	6	7	8
9	10	11	12	13	14	15
16	17	18	19	20	21	22
23	24	25	26	27	28	29
30						

July

Su	M	T	W	Th	F	Sa
	1	2	3	4	5	6
7	8	9	10	11	12	13
14	15	16	17	18	19	20
21	22	23	24	25	26	27
28	29	30	31			

August

Su	M	T	W	Th	F	Sa
				1	2	3
4	5	6	7	8	9	10
11	12	13	14	15	16	17
18	19	20	21	22	23	24
25	26	27	28	29	30	31

September

Su	M	T	W	Th	F	Sa
1	2	3	4	5	6	7
8	9	10	11	12	13	14
15	16	17	18	19	20	21
22	23	24	25	26	27	28
29	30					

October

Su	M	T	W	Th	F	Sa
		1	2	3	4	5
6	7	8	9	10	11	12
13	14	15	16	17	18	19
20	21	22	23	24	25	26
27	28	29	30	31		

November

Su	M	T	W	Th	F	Sa
					1	2
3	4	5	6	7	8	9
10	11	12	13	14	15	16
17	18	19	20	21	22	23
24	25	26	27	28	29	30

December

Su	M	T	W	Th	F	Sa
1	2	3	4	5	6	7
8	9	10	11	12	13	14
15	16	17	18	19	20	21
22	23	24	25	26	27	28
29	30	31				

1. How many months are in a year? ______________________

2. How many days are in a leap year? ______________________

3. What month comes after June? ______________________

4. What day comes before Friday? ______________________

5. How many days are in a normal year? ______________________

Unit 8 - TIME

A clock has 12 numbers and 2 hands. On a clock, the short hand points to the hour and the long hand points to the minute.

Do you see how the arrow pointing to the 12 is longer than the arrow pointing to the 1?

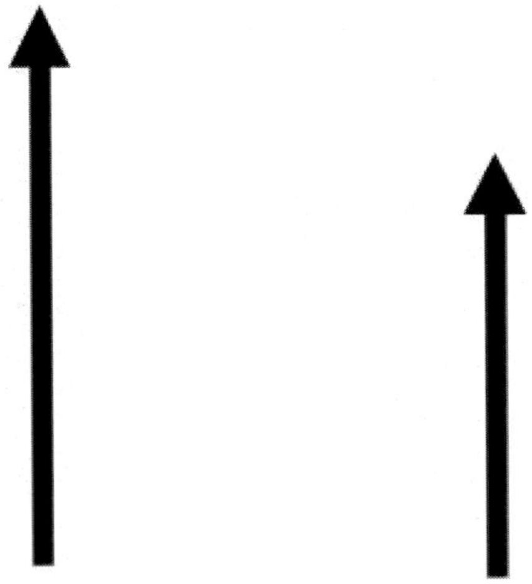

BIG HAND
MINUTE HAND

LITTLE HAND
HOUR HAND

The little hand tells us the hour and the big hand tells us how many minutes after the hour it is right now.

When the big hand points to the 12 and the little hand points to one of the numbers, this tells us the hour.

On this clock, the little hand is point to the 3 and the big hand is pointing to the 12. This means that the time is on the hour. The hour is 3. The time is 3:00.

Name: ____

Write the time shown on each clock.

Remember to write the time as NUMBER : 00.

1.

______ : ______

2.

______ : ______

3.

______ : ______

4.

______ : ______

5.

______ : ______

6.

______ : ______

Name: ______________________________

Write the time shown on each clock.

Remember to write the time as NUMBER : 00.

1.

______ : ______

2.

______ : ______

3.

______ : ______

4.

______ : ______

5.

______ : ______

6.

______ : ______

Name: ______________________________

Draw the hands on the clock to show the time.

1.

3:00

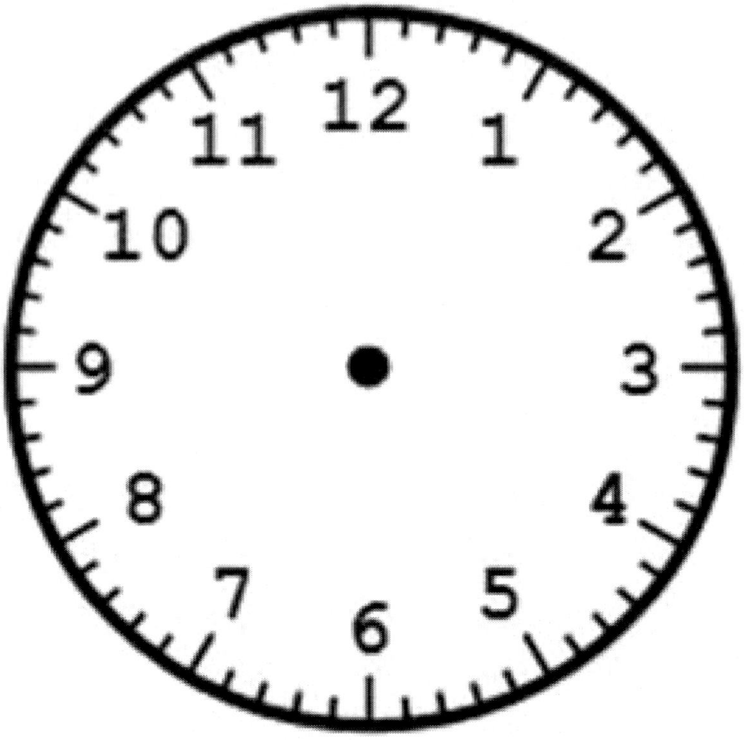

2.

6:00

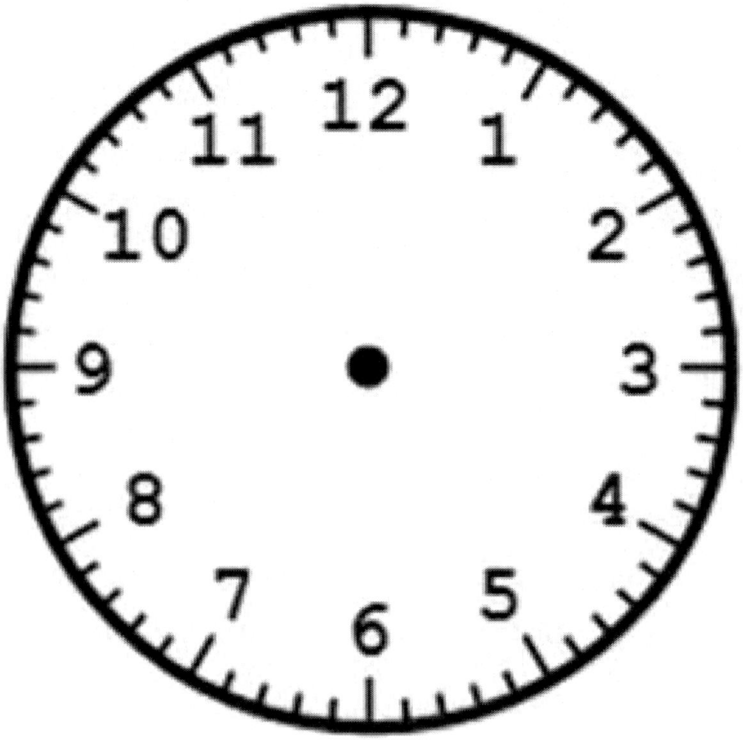

Name: ______________________________

Draw the hands on the clock to show the time.

1.

1:00

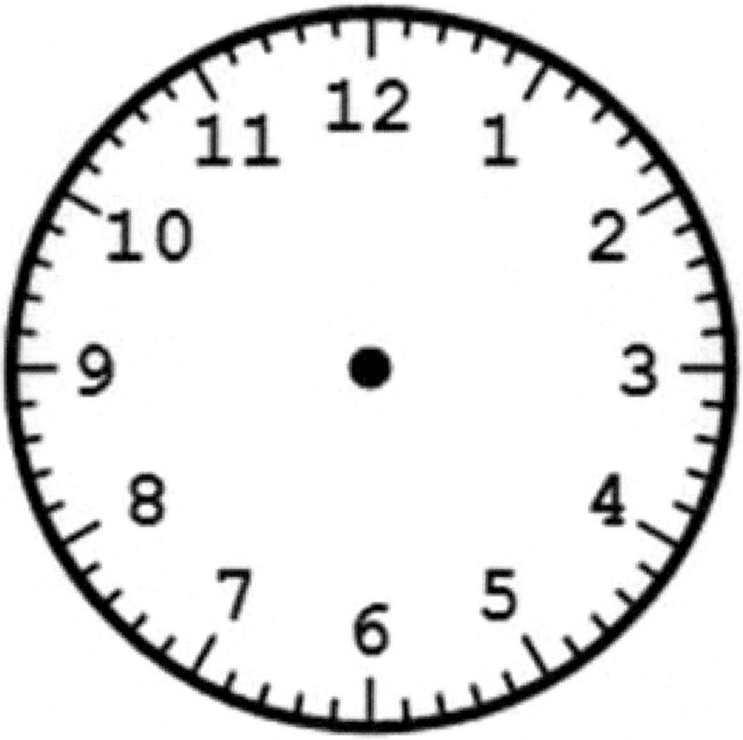

2.

12:00

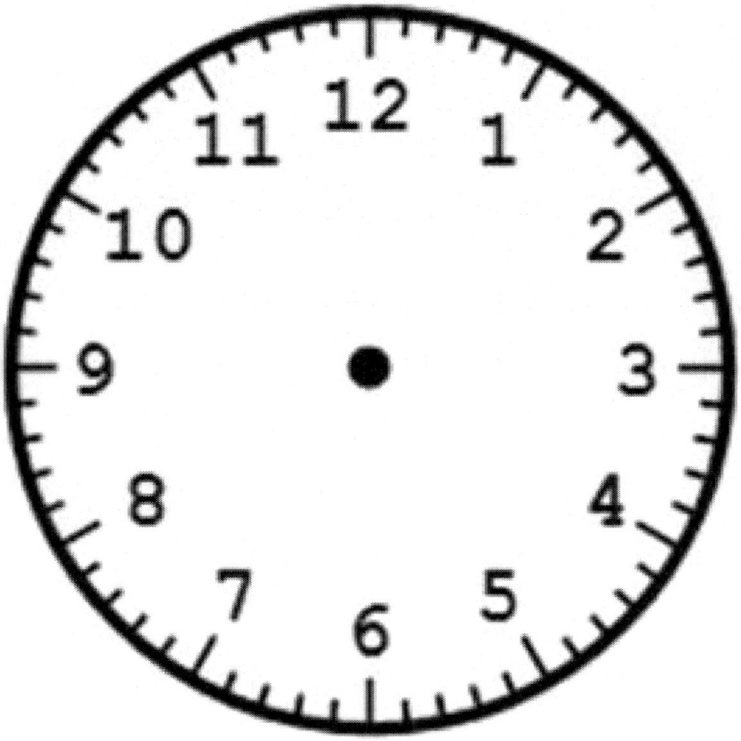

Name: ____________________

Draw the hands on the clock to show the time.

1.

10:00

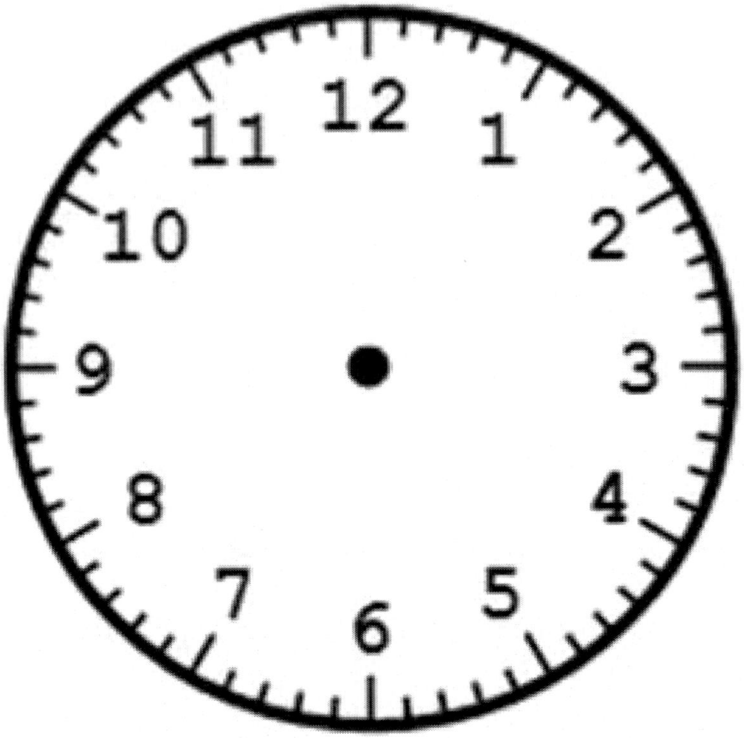

2.

2:00

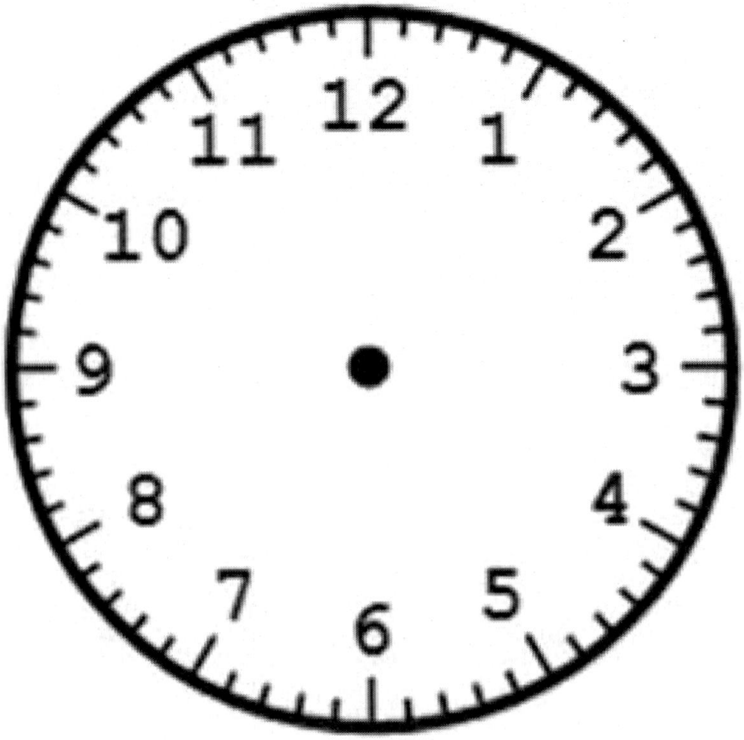

Name: ______________________________

Draw the hands on the clock to show the time.

1.

4:00

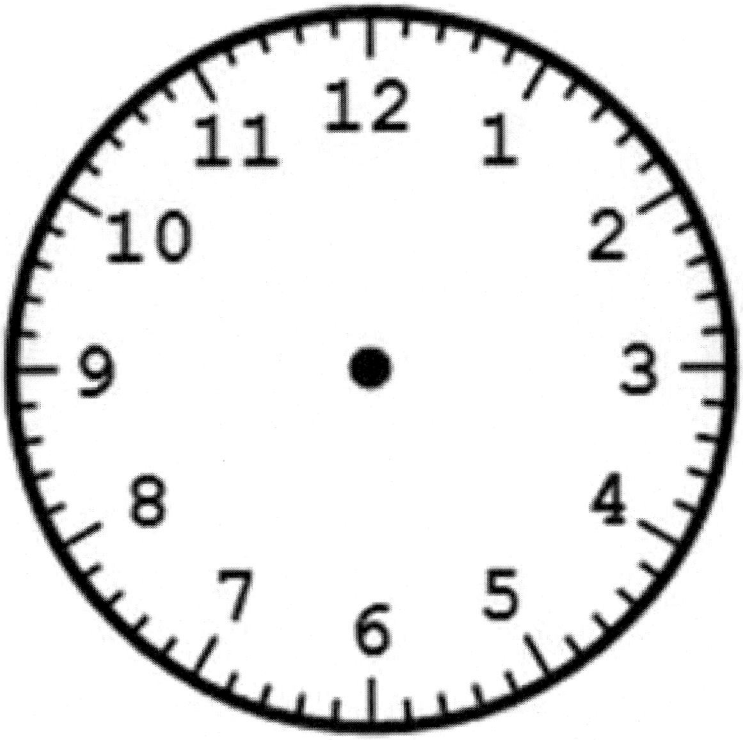

2.

11:00

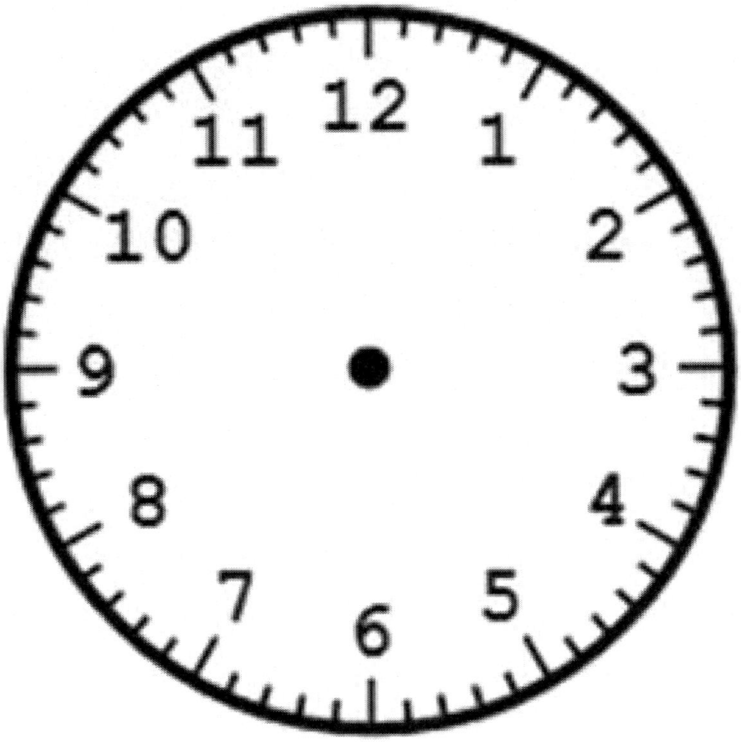

Name: ____

Draw the hands on the clock to show the time.

1.

5:00

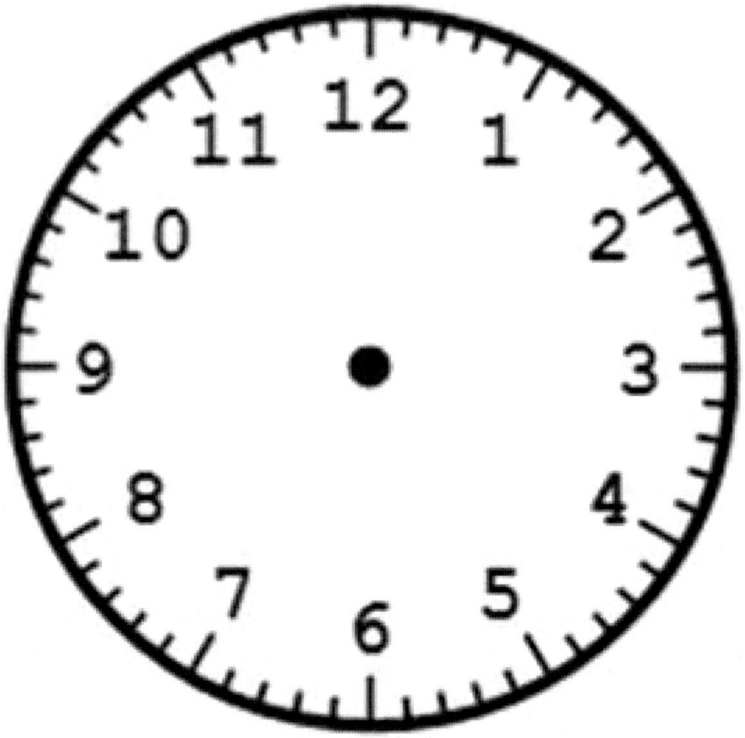

2.

8:00

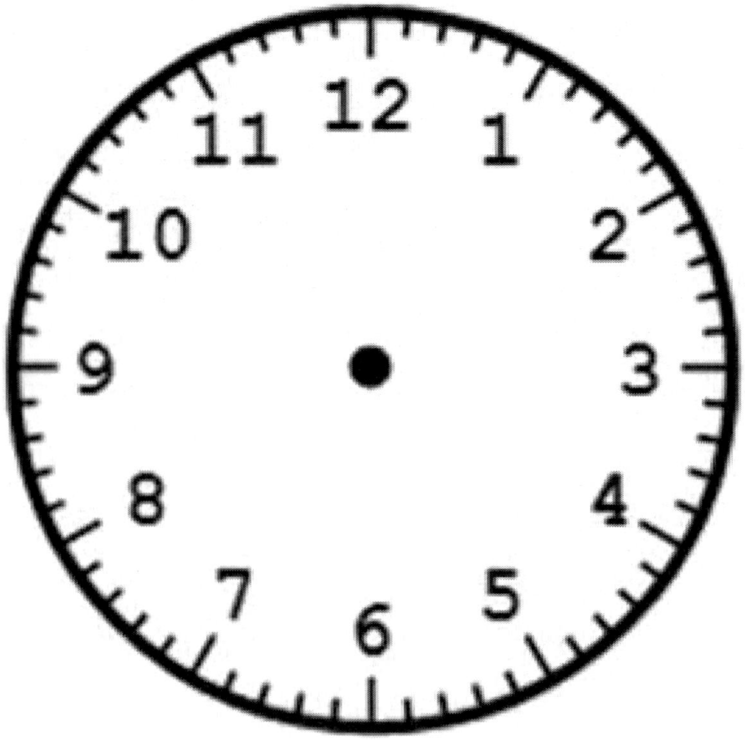

Name: ____

Draw the hands on the clock to show the time.

1.

7:00

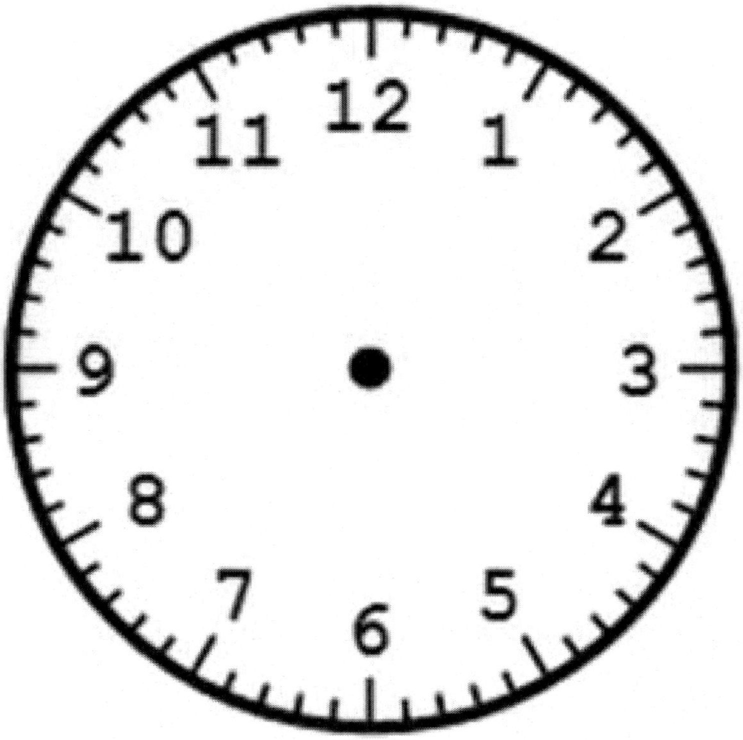

2.

9:00

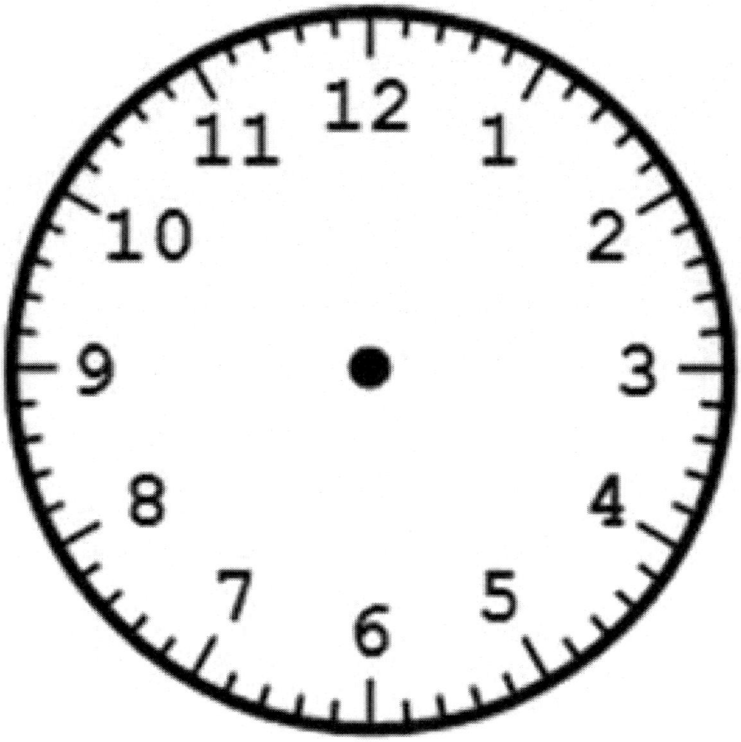

UNIT 9 – FIVE MINUTES

The clock is divided into smaller parts so we can tell the exact time.

Every hour has 60 minutes. Look closely at the clocks in your house. Minutes are marked by tiny lines between the numbers.

If you count all of the lines on the clock, they would equal 60. There are 60 minutes in 1 hour.

The big hand or minute hand points to one of the line places on the clock every minute.

Instead of counting every line to tell time, you can use your skip counting skills to figure out the time faster.

Count the lines between the 5 and the 6, landing on the 6.

Do you see how there are 5 minutes when you jump between the numbers?

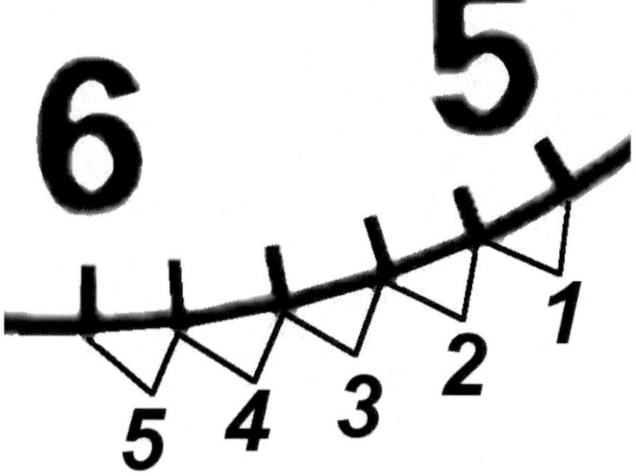

You can use your skip counting skills to count 5 as you land on each number of the clock.

When we count on a clock, we always start at the top. 12 is at the top so we count the 12 as starting at 0. Every time you land on a number, skip count 5 minutes.

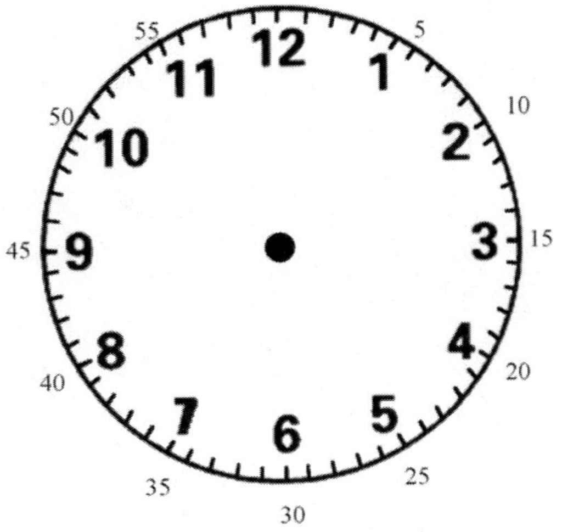

When we say time, we begin with the hour and then add the minutes.

Look at the hour hand. It is pointing at the three.

Look at the minute hand. It is pointing at the one. We know that the one stands for 5 minutes when we skip count.

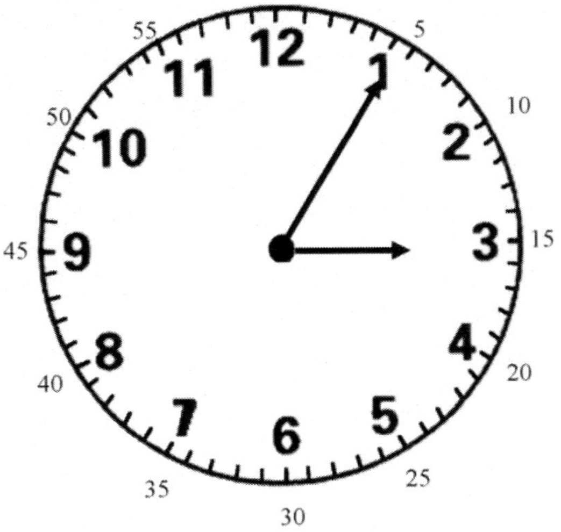

This clock shows 3:05 because the little hand or hour hand is pointing at the 3 and the big hand or minute hand is pointing at the 5 minute mark.

Name: ______________________________

Write the time shown by each clock. Remember, we always start with the hour and then the minutes.

1.

____:____

2.

____:____

3.

____:____

4..

____:____

5.

____:____

6.

____:____

Name: ____

Write the time shown by each clock. Remember, we always start with the hour and then the minutes.

1.

_____ : _____

2.

_____ : _____

3.

_____ : _____

4.

_____ : _____

5.

_____ : _____

6.

_____ : _____

Name: ____

Write the time shown by each clock. Remember, we always start with the hour and then the minutes.

1.

_____:_____

2.

_____:_____

3.

_____:_____

4.

_____:_____

5.

_____:_____

6.

_____:_____

Name: ______________________________

Write the time shown by each clock. Remember, we always start with the hour and then the minutes.

1. ______:______

2. ______:______

3. ______:______

4. ______:______

5. ______:______

6. ______:______

Name: ____

Write the time shown by each clock. Remember, we always start with the hour and then the minutes.

1.

______ : ______

2.

______ : ______

3.

______ : ______

4.

______ : ______

5.

______ : ______

6.

______ : ______

Name: ______________________________

Write the time shown by each clock. Remember, we always start with the hour and then the minutes.

1.

_____ : _____

2.

_____ : _____

3.

_____ : _____

4.

_____ : _____

5.

_____ : _____

6.

_____ : _____

Name: ____

Write the time shown by each clock. Remember, we always start with the hour and then the minutes.

1.

_______:_______

2.

_______:_______

3.

_______:_______

4.

_______:_______

5.

_______:_______

6.

_______:_______

Name: ____________________

Draw the hands on the clock to show the time.

4:40

3:10

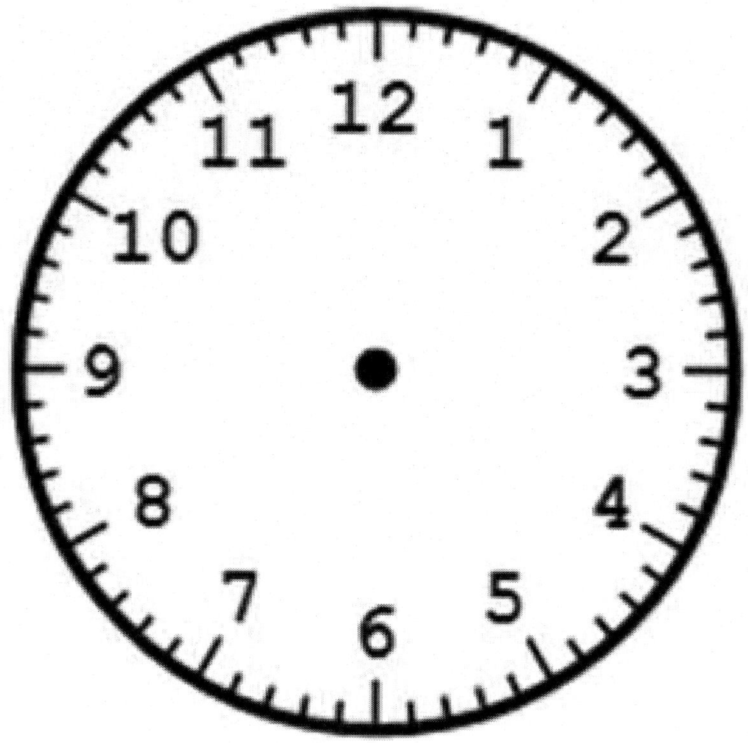

Name: ______________________________

Draw the hands on the clock to show the time.

8:05

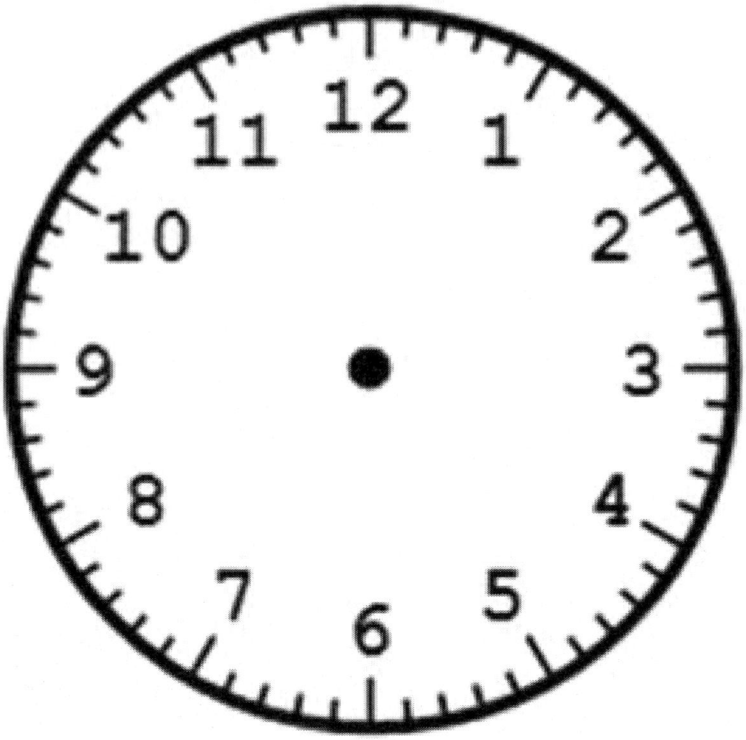

12:55

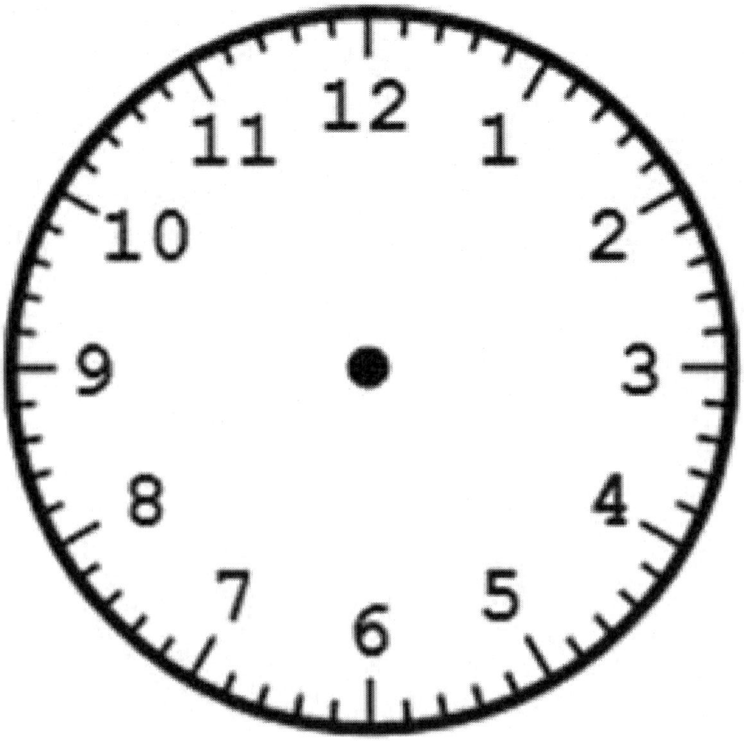

Name: ____

Draw the hands on the clock to show the time.

8:45

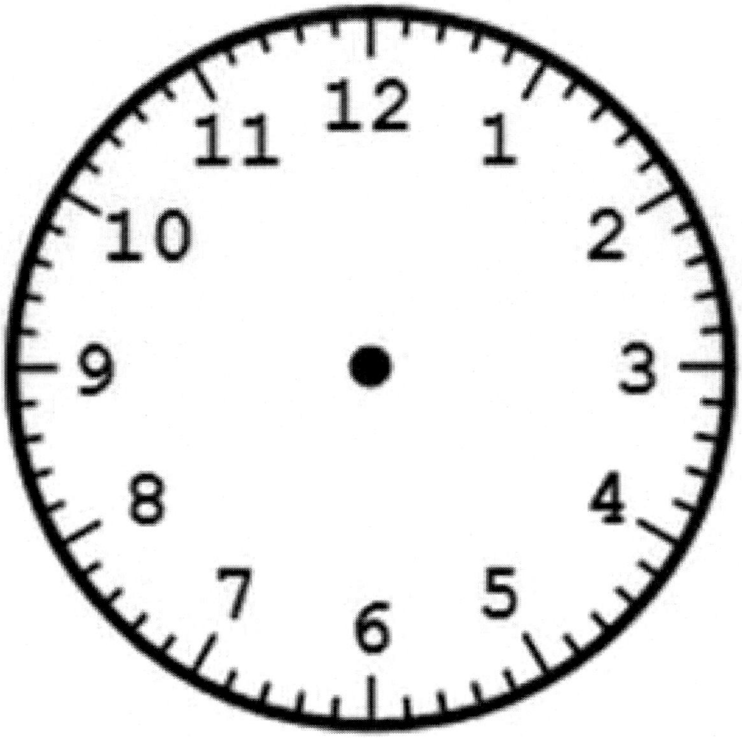

12:25

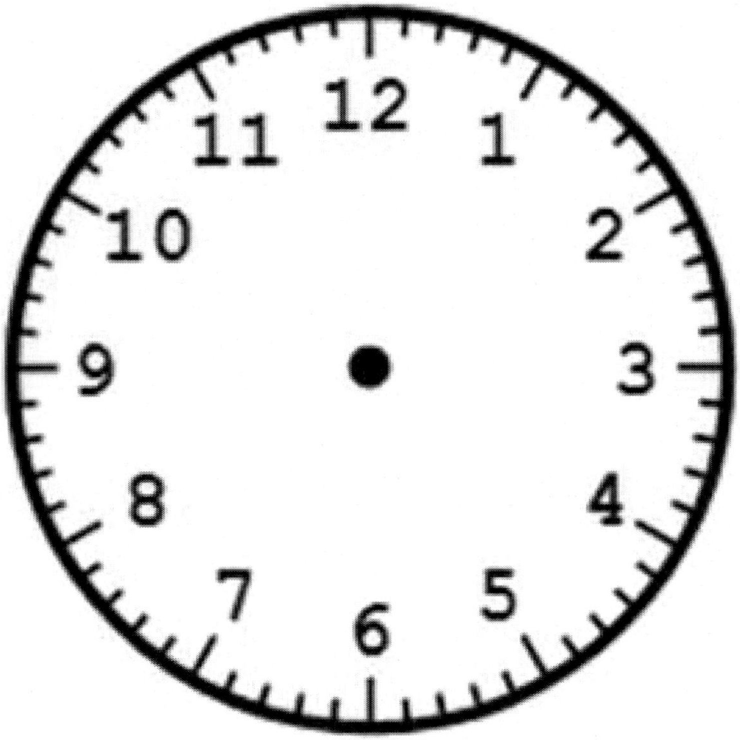

Name: ____________________

Draw the hands on the clock to show the time.

7:35

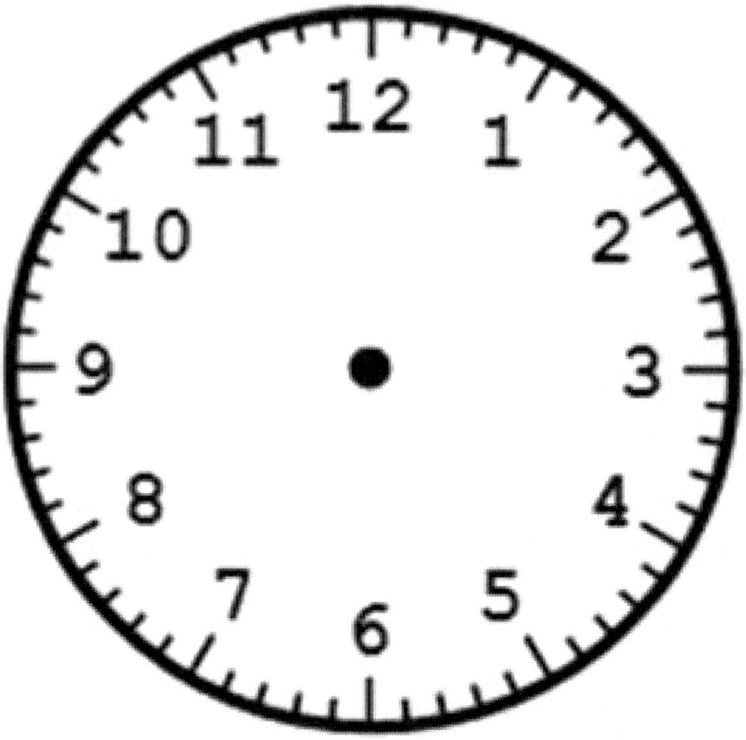

2:15

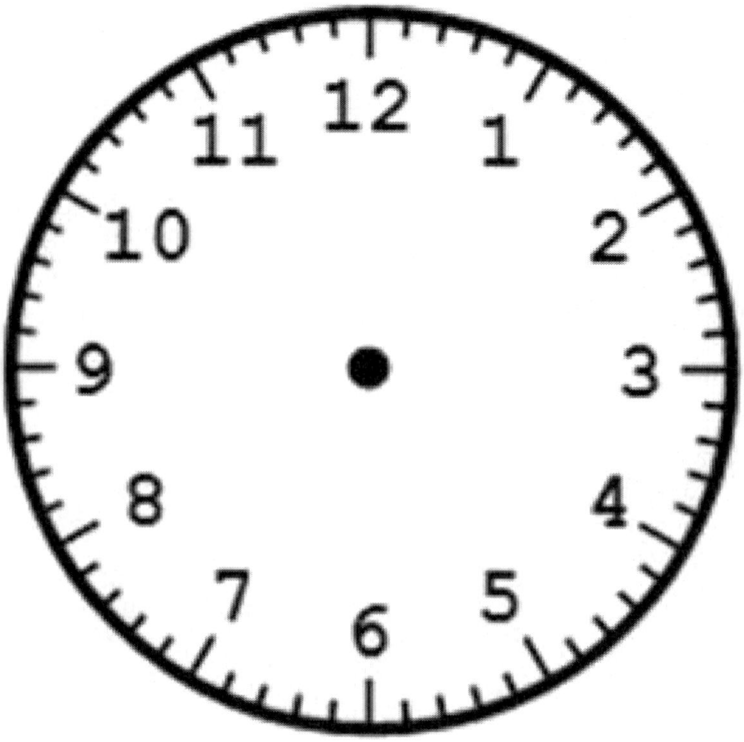

Name: ____

Draw the hands on the clock to show the time.

6:20

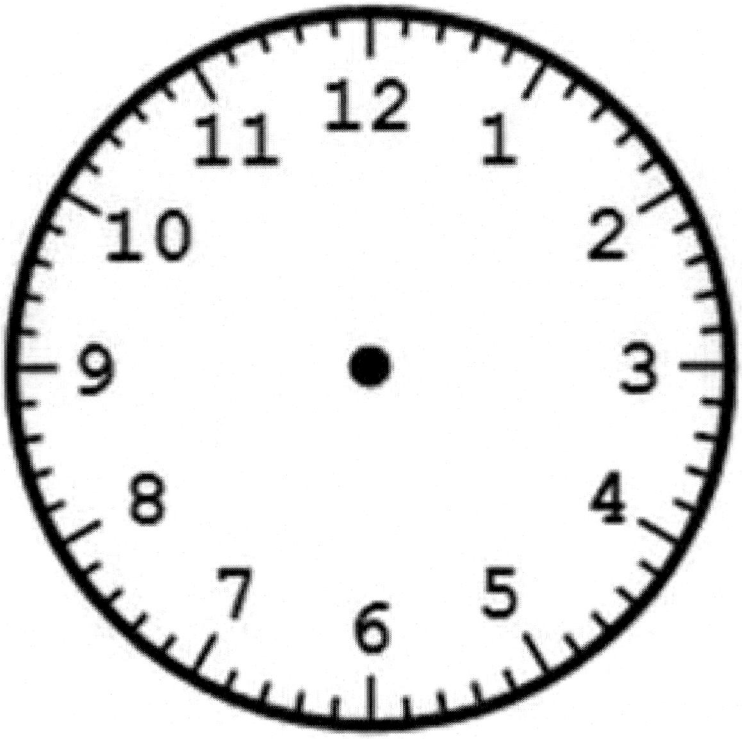

10:30

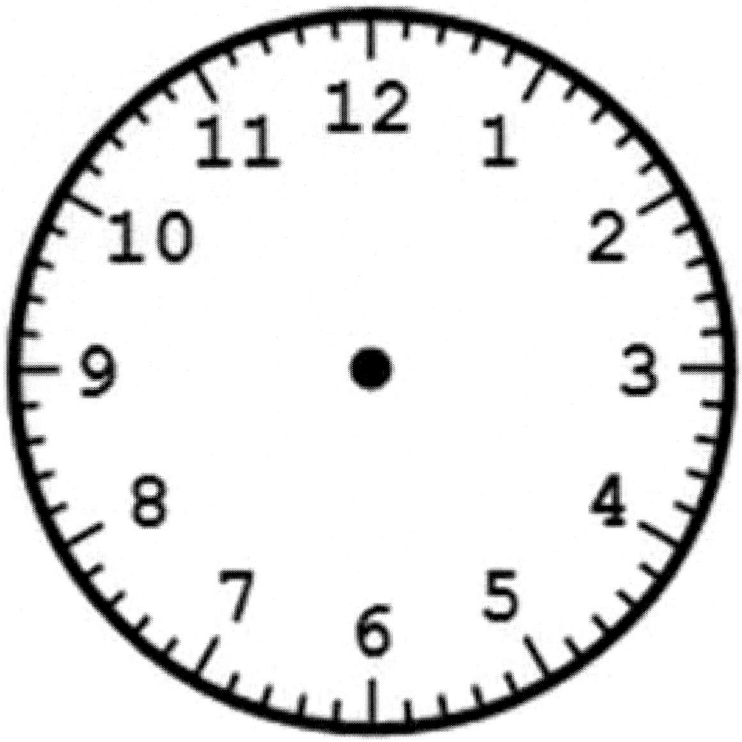

Name: ______________________________

Draw the hands on the clock to show the time.

9:50

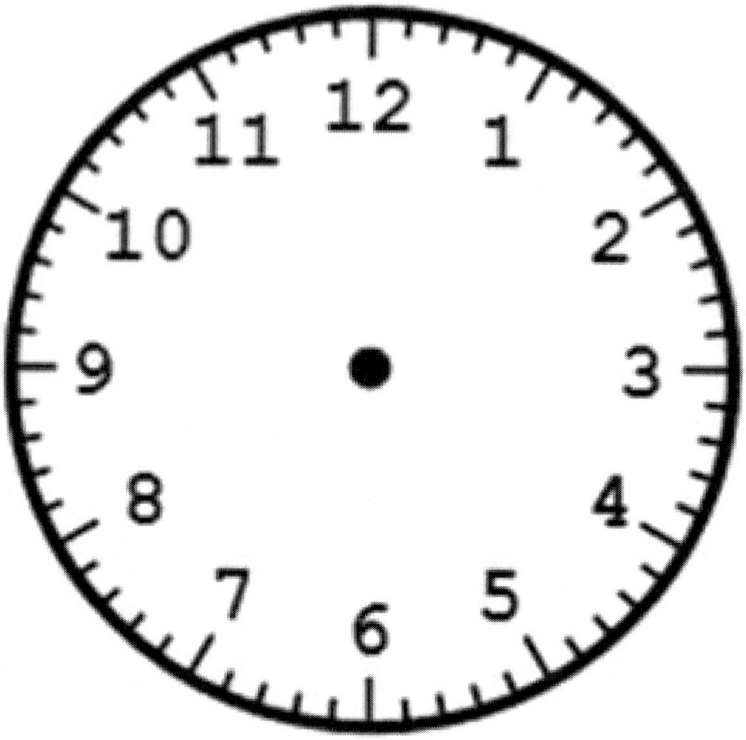

1:35

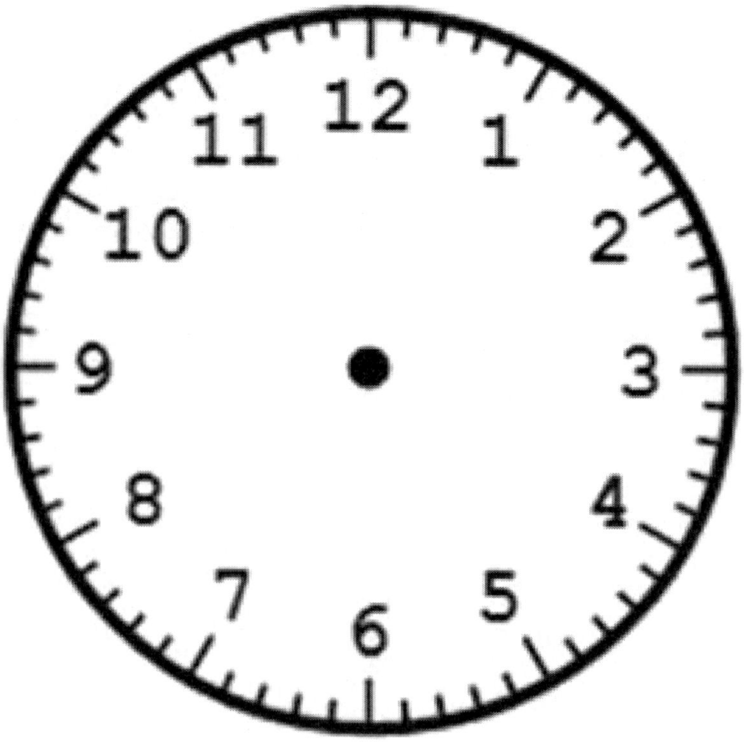

Name: ______________________________

Draw the hands on the clock to show the time.

5:10

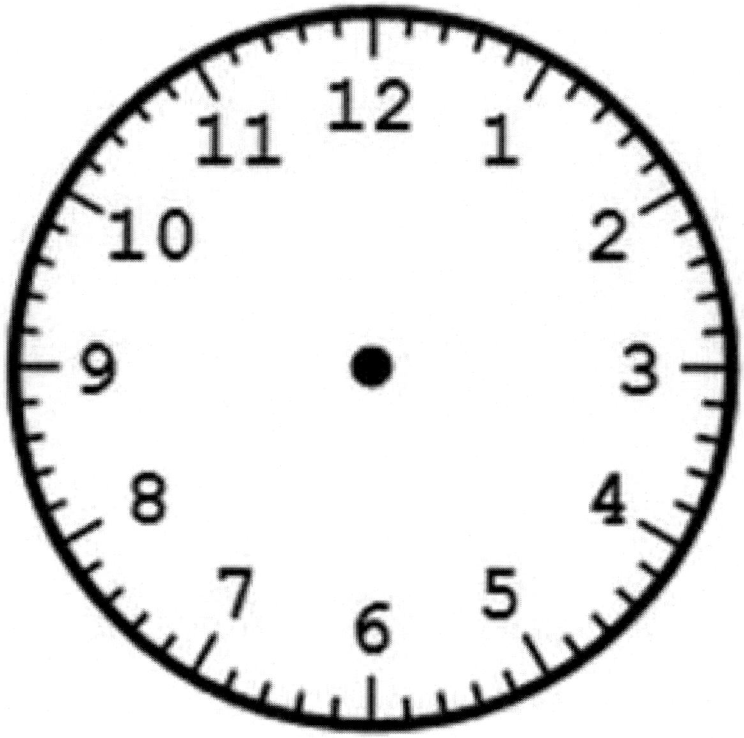

11:25

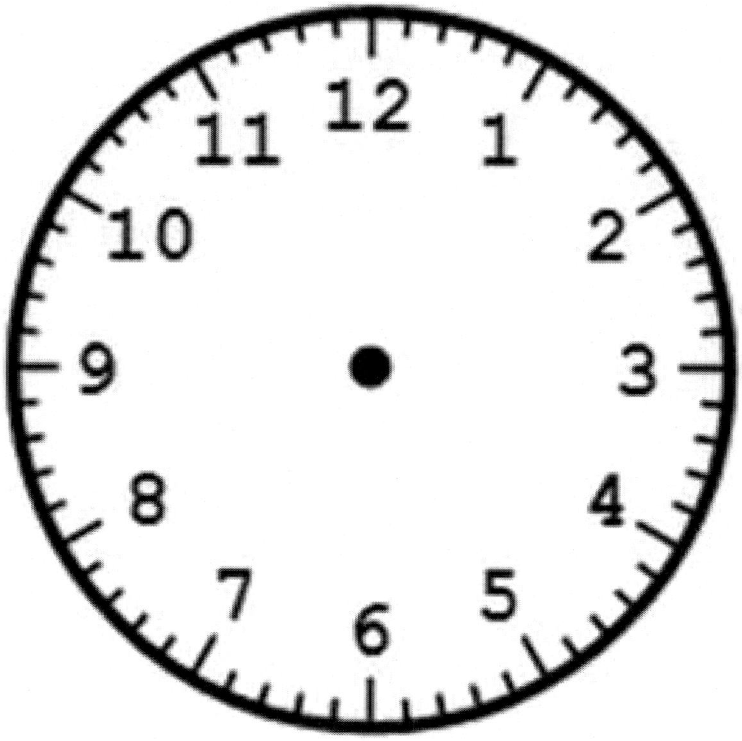

Name: ____________________

Draw the hands on the clock to show the time.

10:45

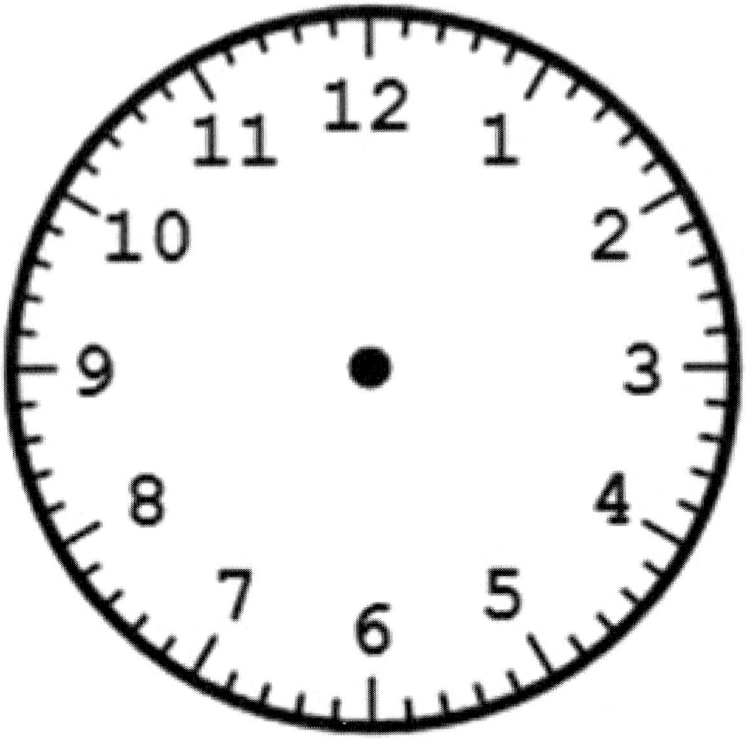

3:20

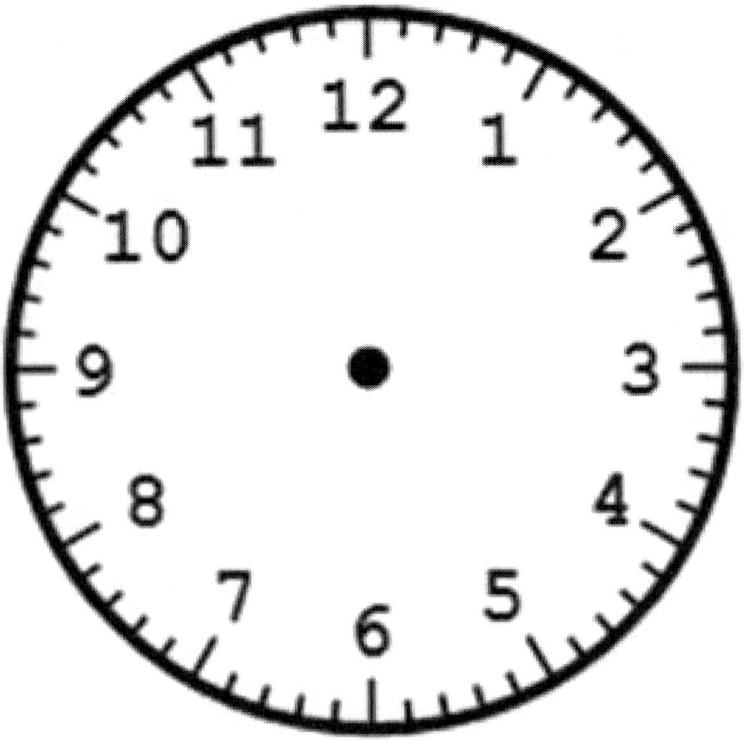

UNIT 10 – HALF HOUR

You can make telling time easier by using your fraction skills. Divide the clock in 1/2.

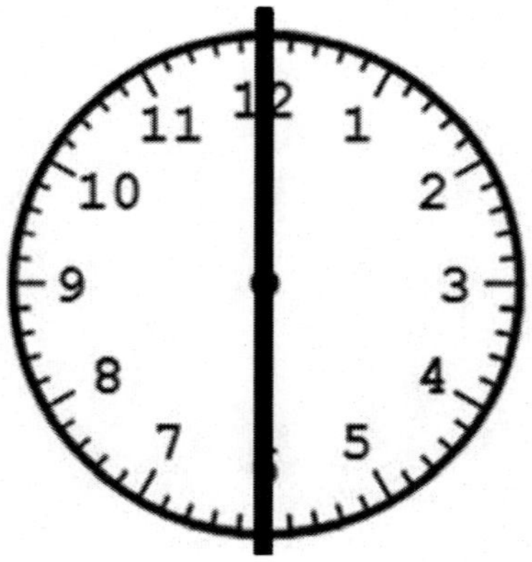

The number 6 is 1/2 way around the clock.

Since the whole clock is an hour, when the minute hand divides the clock in 1/2 we call this the half-hour.

When the minute hand points at the 6, it is 1/2 past the hour.

half past 1

When hour hand points to the 1 and the minute hand is 1/2 way around the clock. We say it is half past 1.

You can also skip count to the minute hand. It is pointing at the number 6.

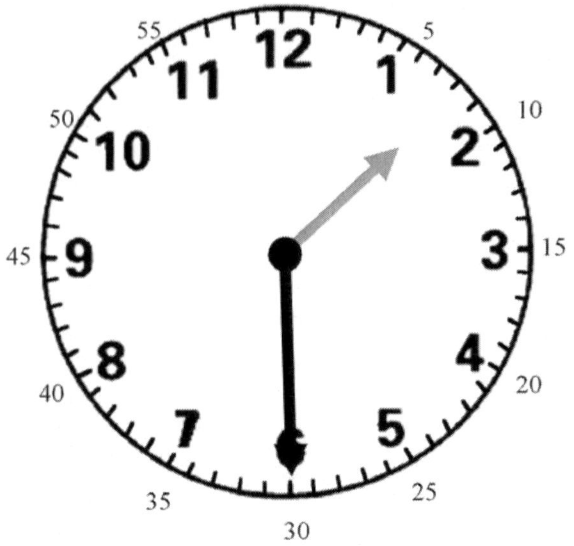

Skip count to the number 6. The number 6 represents 30 minutes. The hour is 1 and the minutes are 30.

We can say 1:30.

We can say half past 1.

Both 1:30 and half past 1 mean the same thing.

Name: ____

Write the time shown by each clock. Write the time using numbers and using half past.

1.

______ : ______

____

2.

______ : ______

____

3.

______ : ______

____

4.

______ : ______

____

5.

______ : ______

____

6.

______ : ______

____

Name: ____

Write the time shown by each clock. Write the time using numbers and using half past.

1.

______:______

____

2.

______:______

____

3.

______:______

____

4.

______:______

____

5.

______:______

____

6.

______:______

____

Name: ____

Write the time shown by each clock. Write the time using numbers and using half past.

1.

_______ : _______

____

2.

_______ : _______

____

3.

_______ : _______

____

4.

_______ : _______

____

5.

_______ : _______

____

6.

_______ : _______

____

Name: ____

Draw the hands on the clock to show the time.

1:30

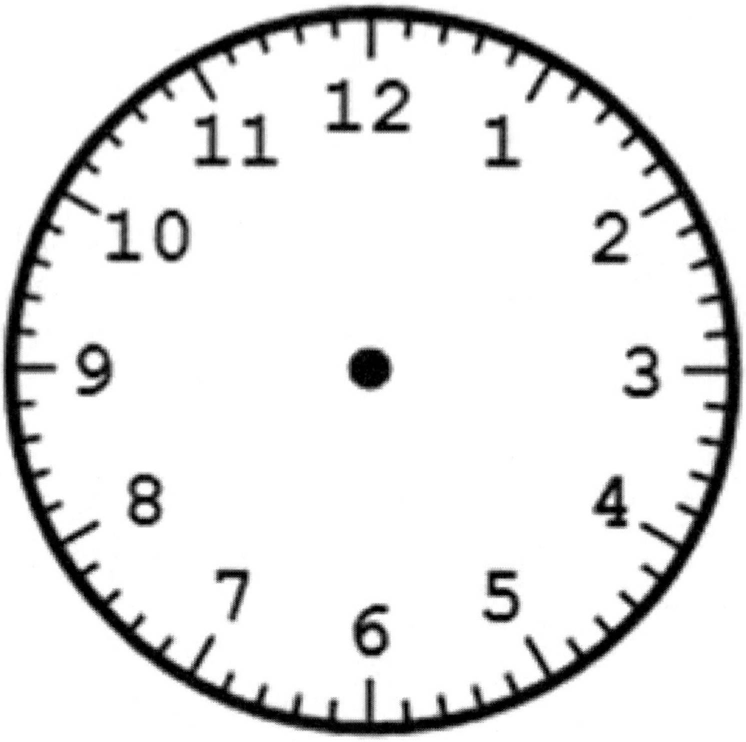

half past 3

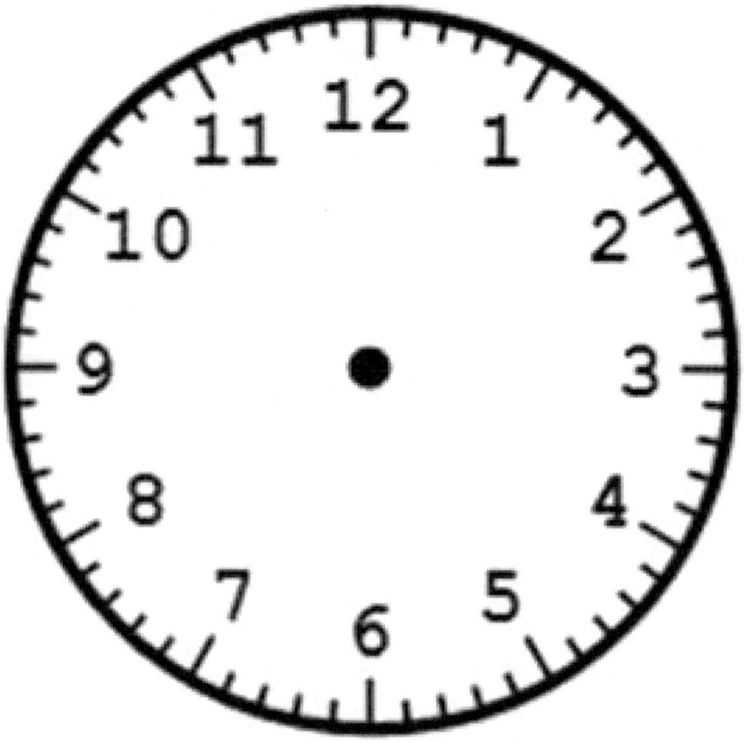

Name: ______________________________

Draw the hands on the clock to show the time.

half past 9

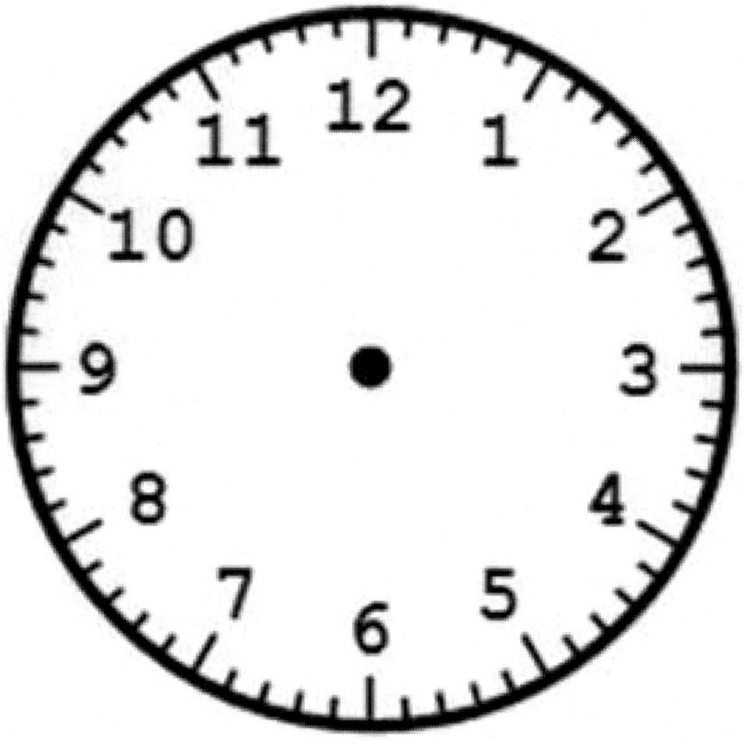

12:30

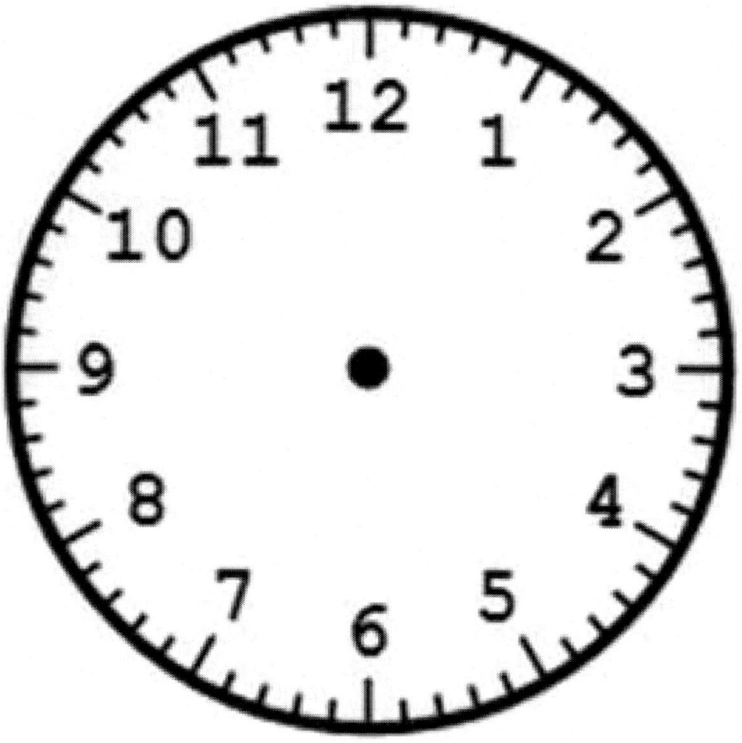

Name: ______________________________

Draw the hands on the clock to show the time.

7:30

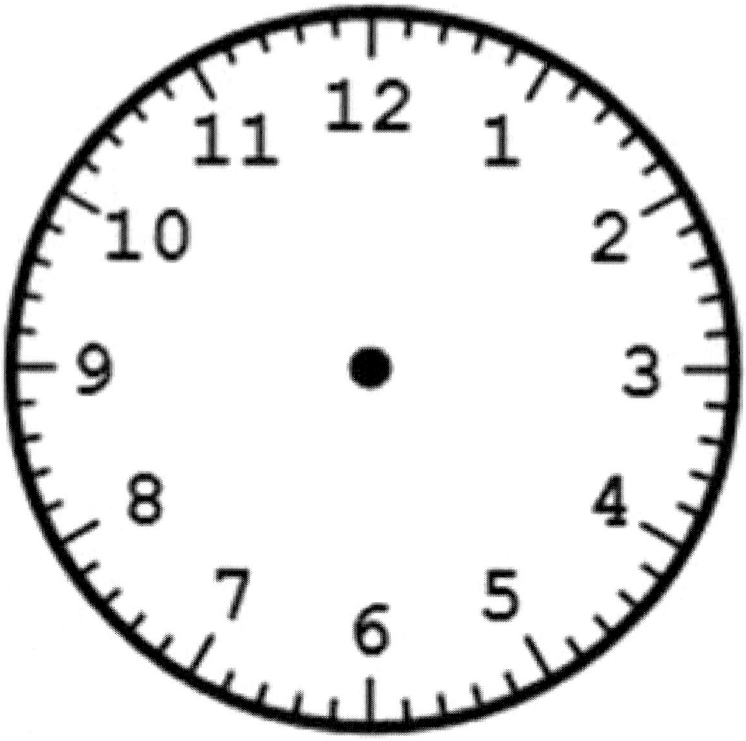

10:30

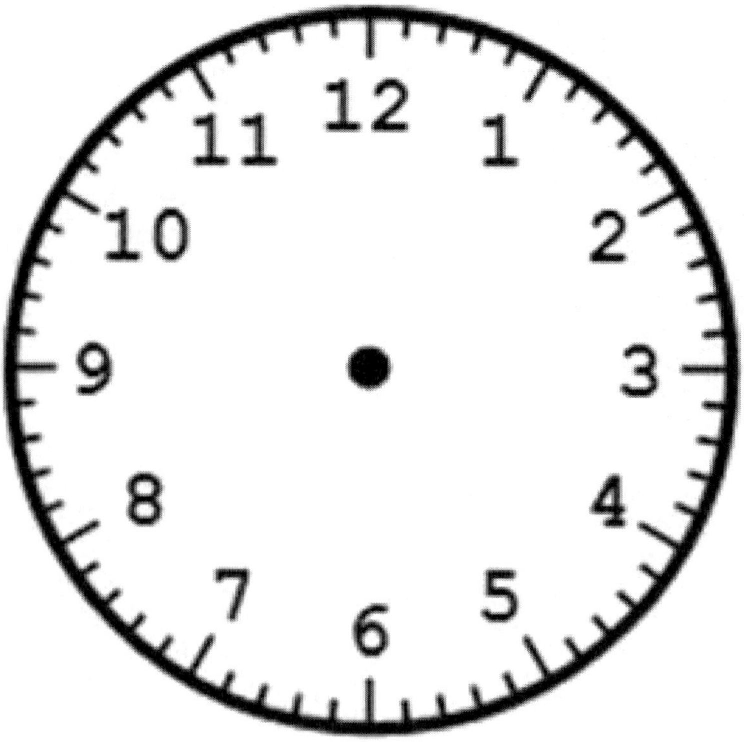

Name: ____________________

Draw the hands on the clock to show the time.

half past 12

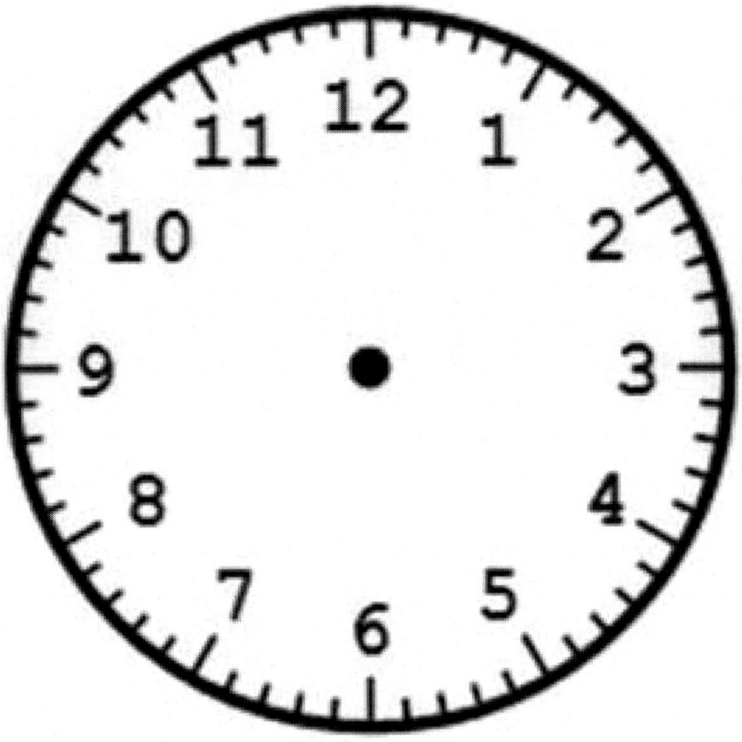

half past 4

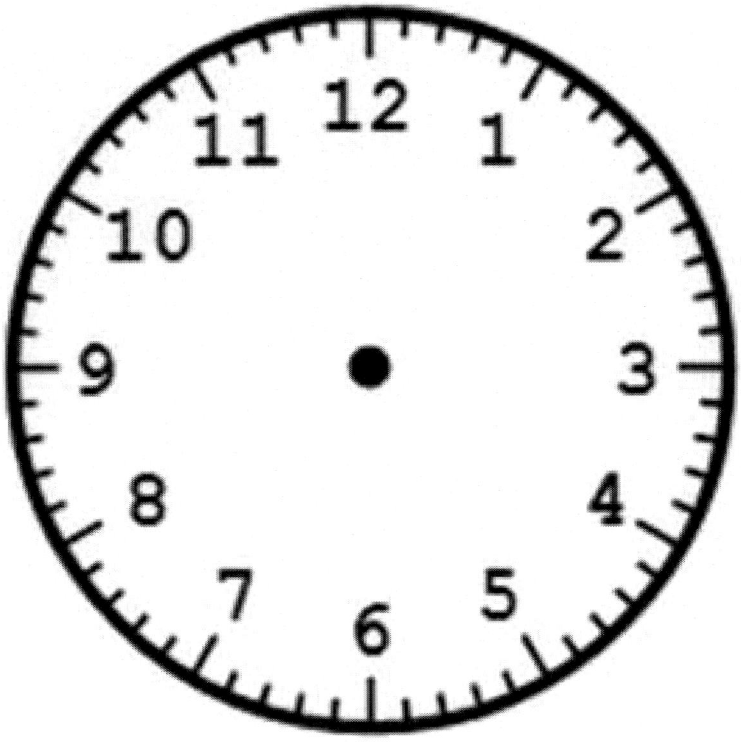

Name: ______________________________

Draw the hands on the clock to show the time.

half past 1

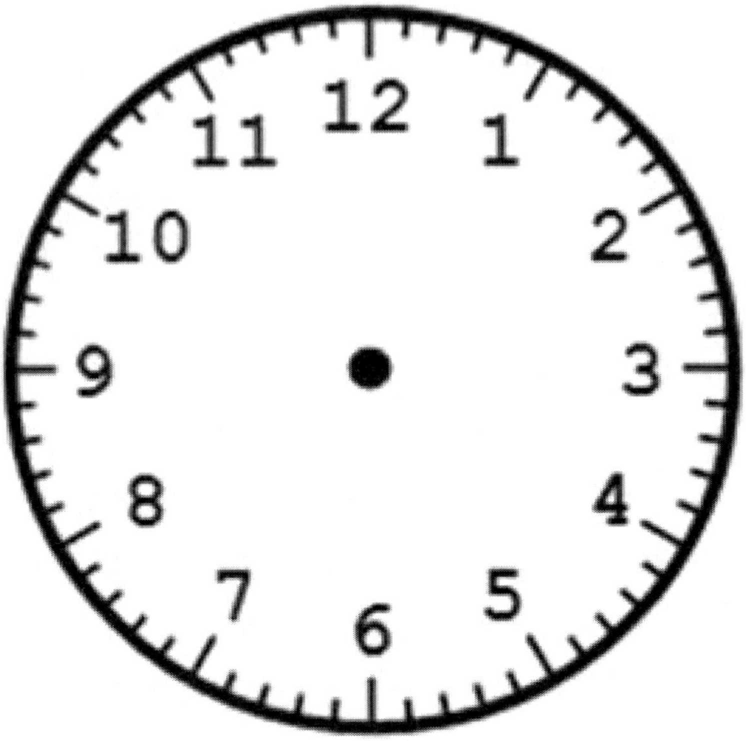

5:30

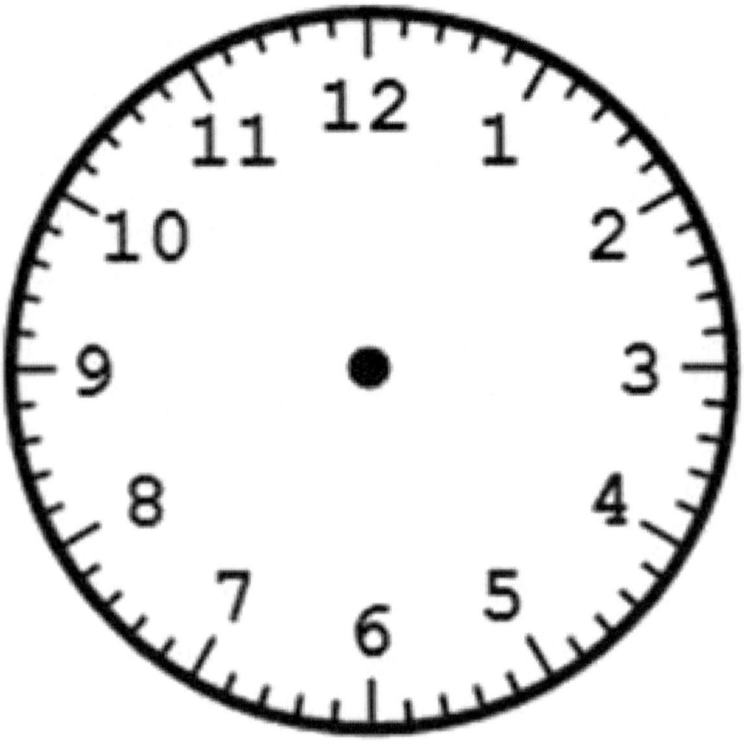

Name: ____

Draw the hands on the clock to show the time.

2:30

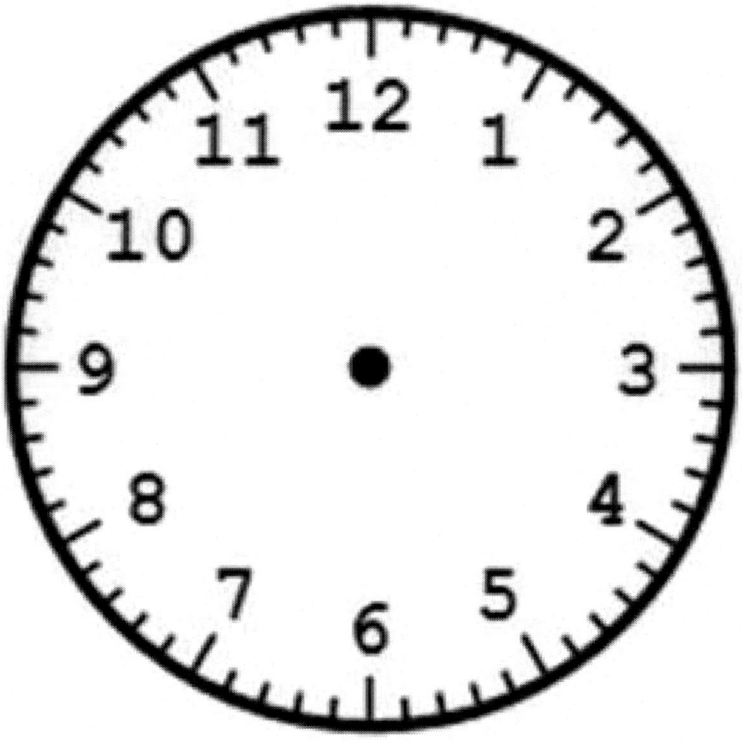

5:30

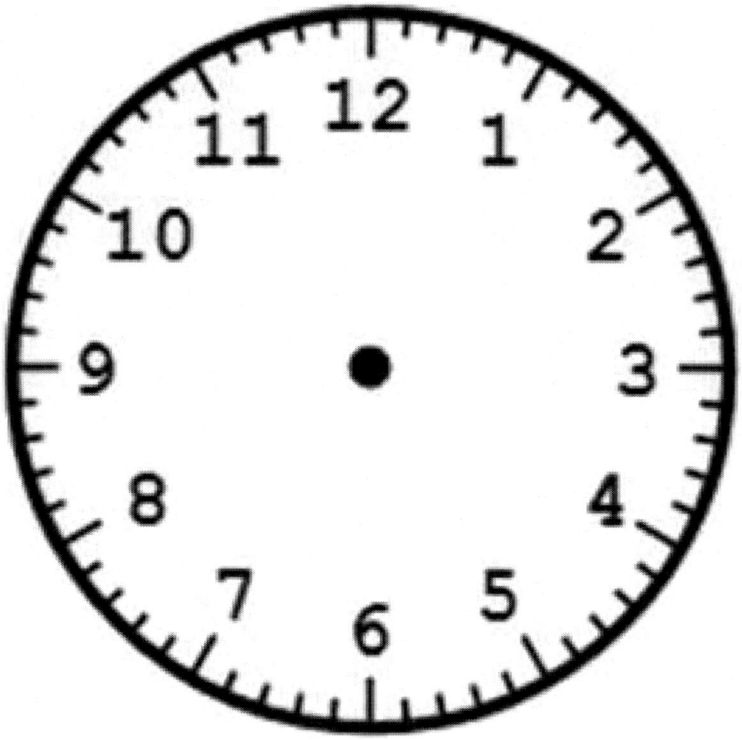

Name: ______________________________

Draw the hands on the clock to show the time.

half past 5

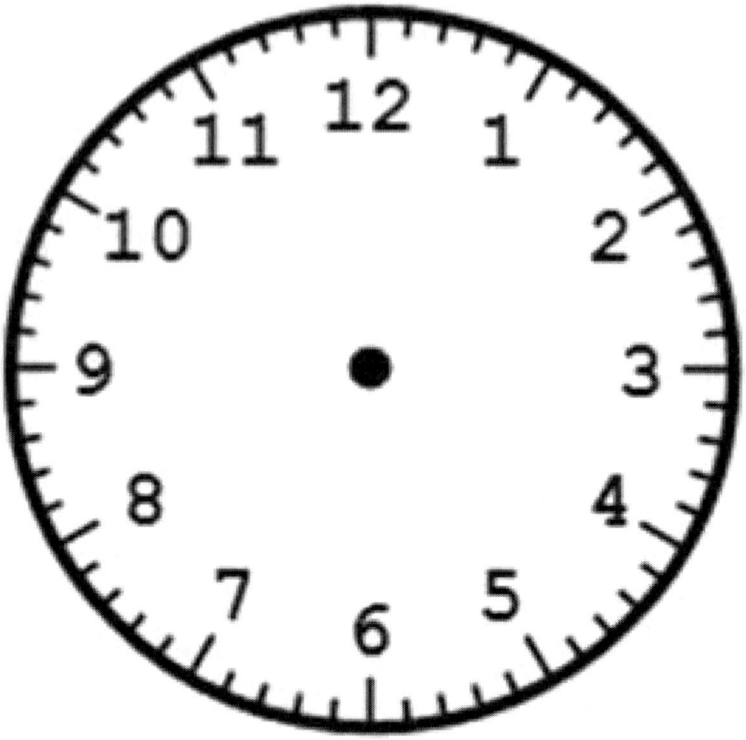

half past 11

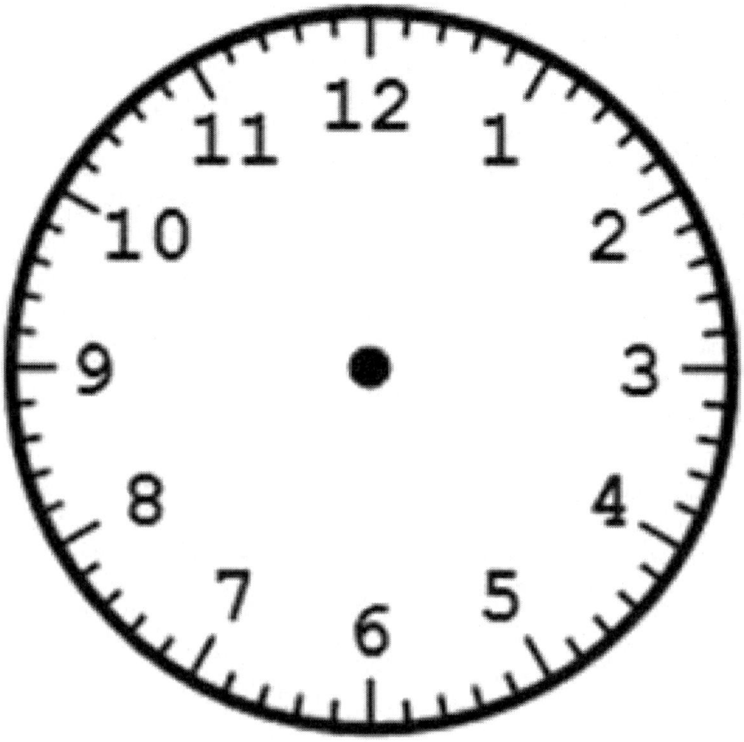

Name: ____________________

Draw the hands on the clock to show the time.

half past 6

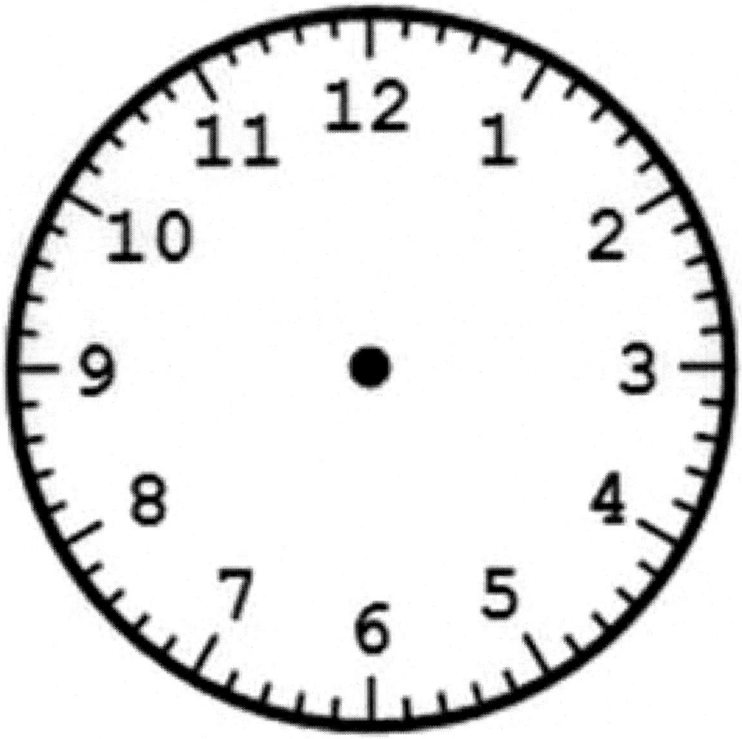

3:30

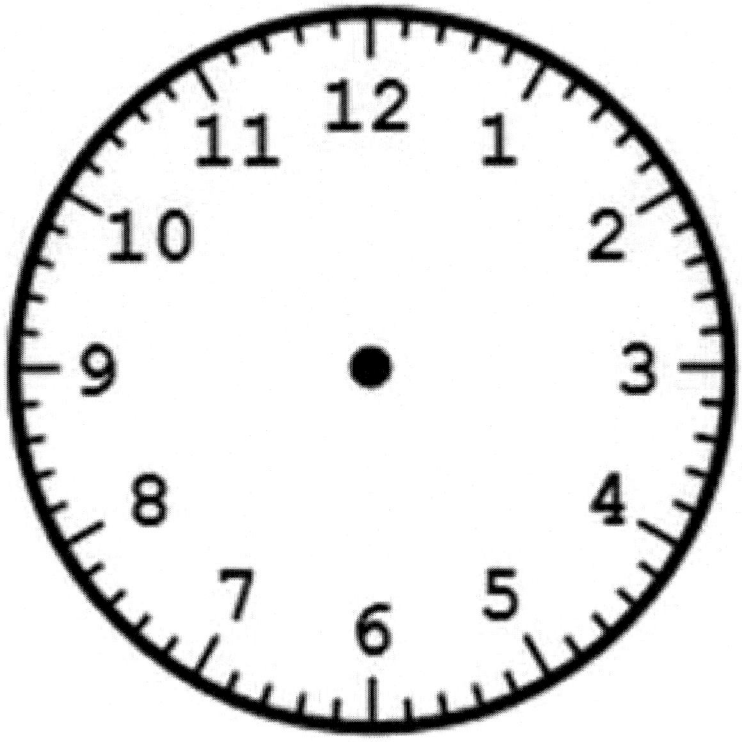

Name: ____

Draw the hands on the clock to show the time.

half past 8

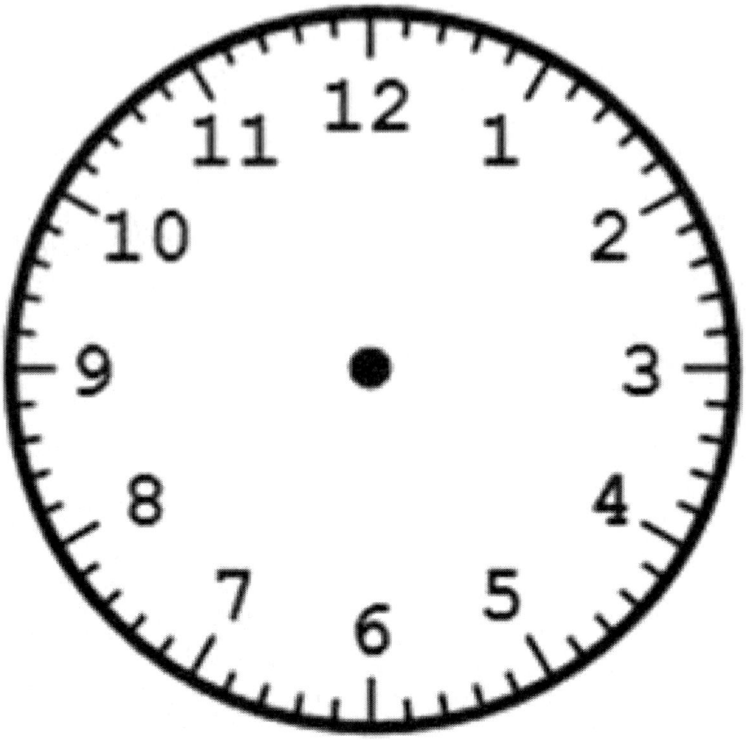

4:30

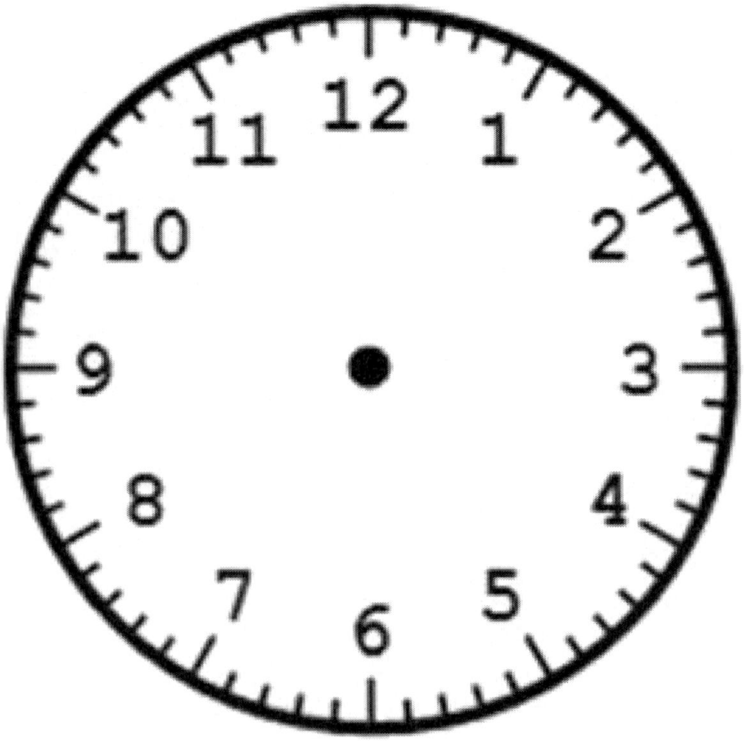

Name: ____

Draw the hands on the clock to show the time.

8:30

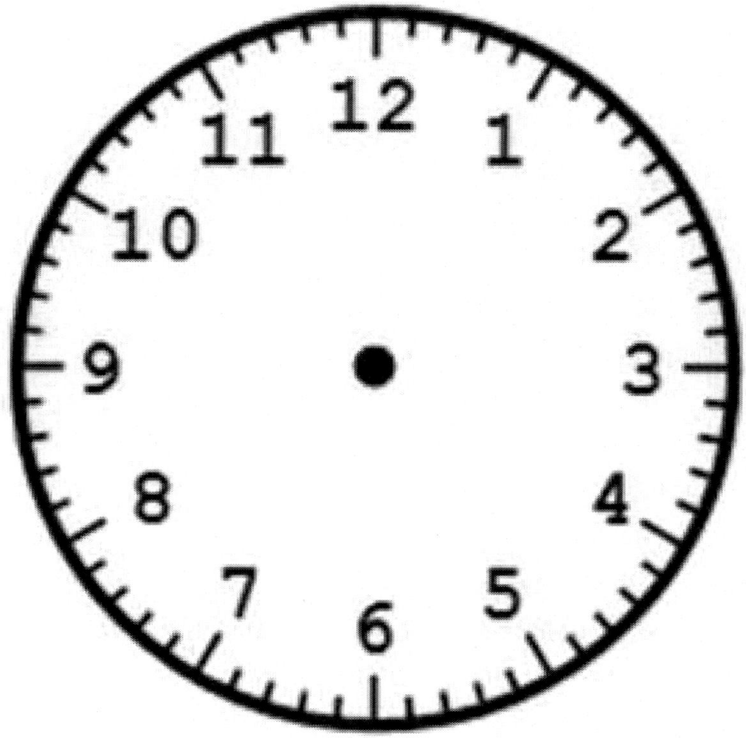

half past 2

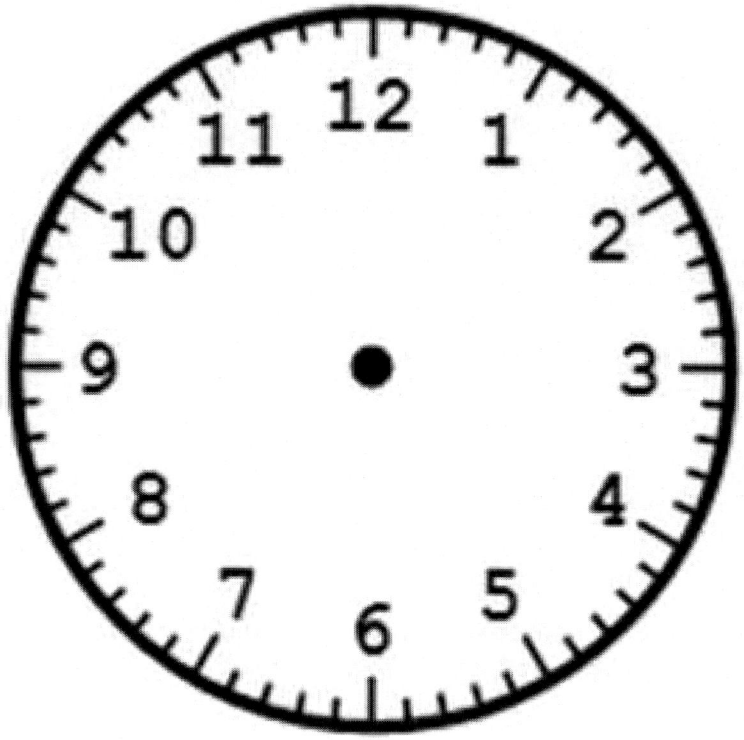

UNIT 11 – QUARTER HOUR

You can use your fraction skills to divide the clock again, making telling time even easier.

You saw how the clock can be divided in 1/2.

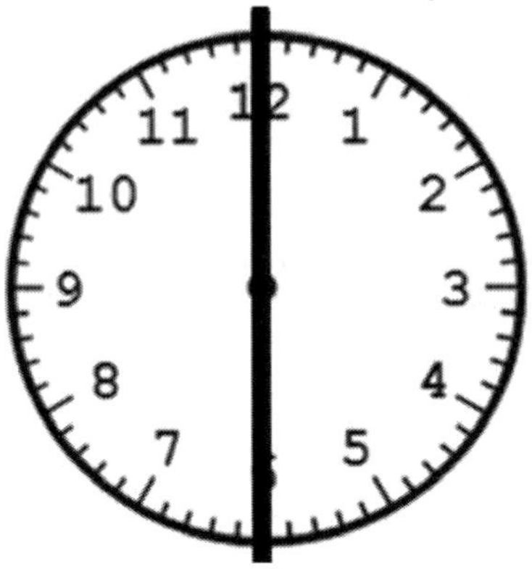

If you divide it again, it is in 1/4 sections.

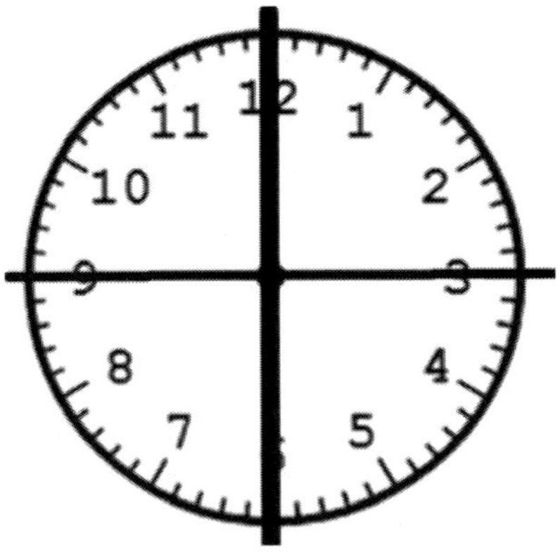

The number 3 is 1/4 of the way around the clock.

The time is a quarter after 3.

You can also skip count to the minute hand. The big hand is pointing at the 3.

Skip count to the number 3. The number 3 represents 15 minutes. The hour is 3 and the minutes are 15.

We can say a quarter after 3.

We can say 3:15.

Both a quarter after 3 and 3:15 mean the same thing.

When we divide the clock into a fraction, the line goes through the 9 too.

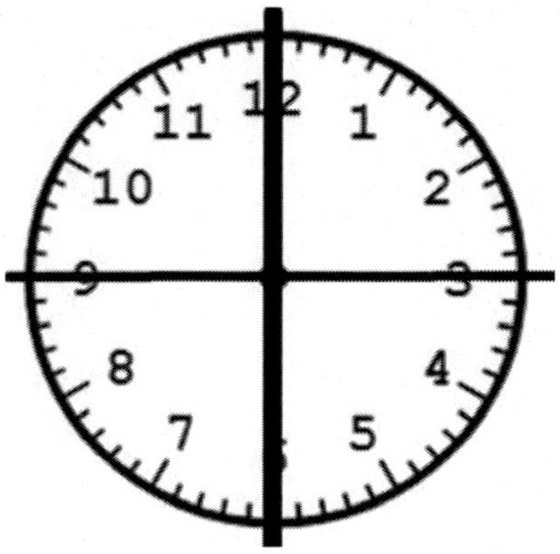

The number 9 is 3/4 of the way around the clock. It would be confusing to say three quarters after 3 so we say a quarter till 4. We say a quarter till 4 because the area between the number 9 and the 12 represents 1/4 of the clock.

You can also skip count to the minute hand. The big hand is pointing at the 9.

Skip count to the number 9. The number 9 represents 35 minutes. The hour is 3 and the minutes are 45.

We can say a quarter till 4.

We can say 3:45.

Both a quarter till 4 and 3:45 mean the same thing.

Name: ______________________________

Write the time shown by each clock. Write the time using numbers and using a quarter after or a quarter till.

1.

_____:_____

____

2.

_____:_____

____

3.

_____:_____

____

4.

_____:_____

____

5.

_____:_____

____

6.

_____:_____

____

Name: ____

Write the time shown by each clock. Write the time using numbers and using a quarter after or a quarter till.

1.

______ : ______

____

2.

______ : ______

____

3.

______ : ______

____

4.

______ : ______

____

5.

______ : ______

____

6.

______ : ______

____

Name: ____

Write the time shown by each clock. Write the time using numbers and using a quarter after or a quarter till.

1.

______:______

____

2.

______:______

____

3.

______:______

____

4.

______:______

____

5.

______:______

____

6.

______:______

____

Name: ______________________________

Write the time shown by each clock. Write the time using numbers and using a quarter after or a quarter till.

1.

_____ : _____

____

2.

_____ : _____

____

3.

_____ : _____

____

4.

_____ : _____

____

5.

_____ : _____

____

6.

_____ : _____

____

Name: ____________________

Draw the hands on each clock to show the time.

8:45

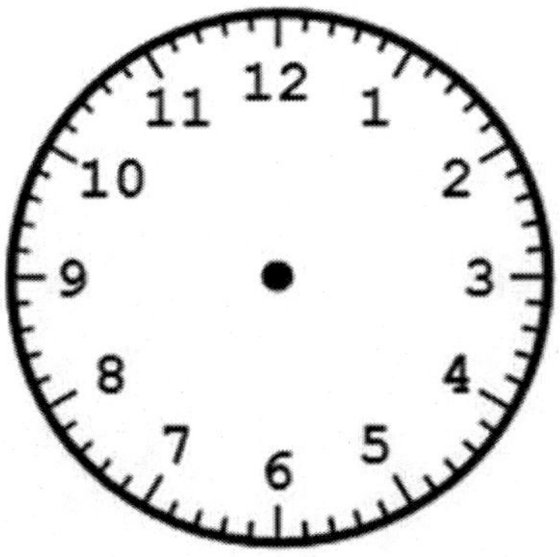

a quarter till 2

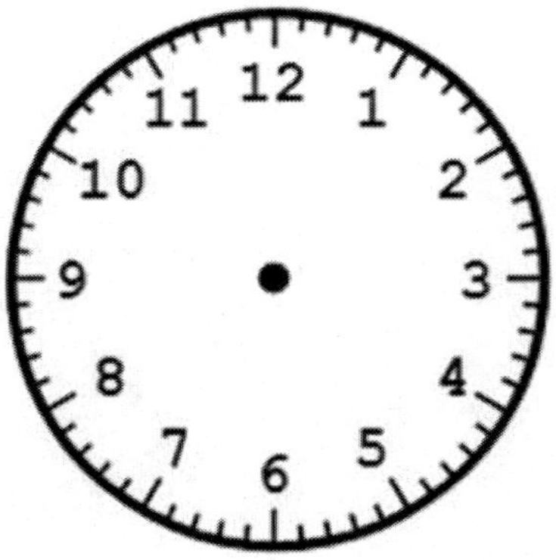

a quarter after 12

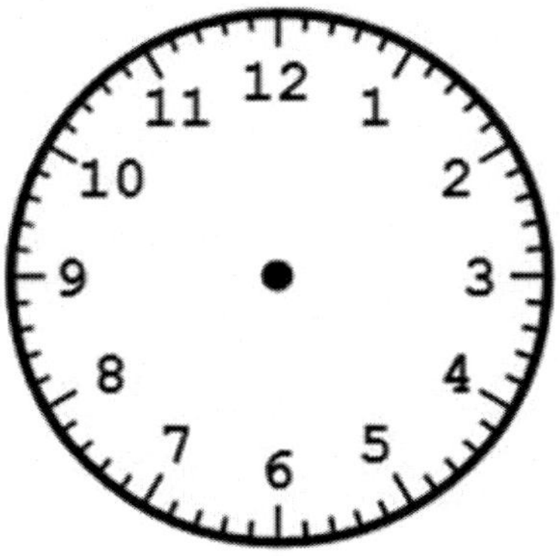

1:15

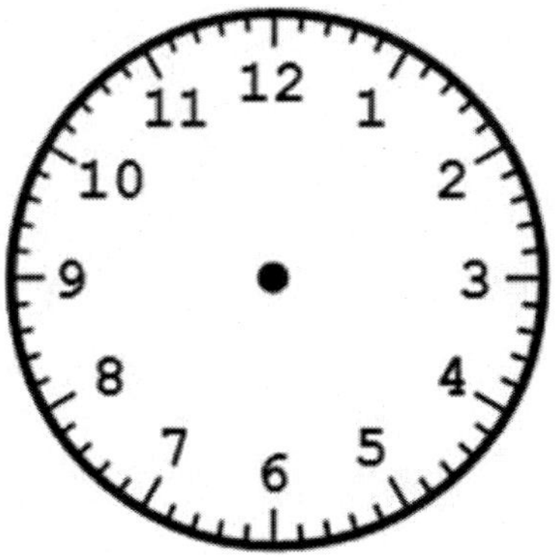

a quarter till 12

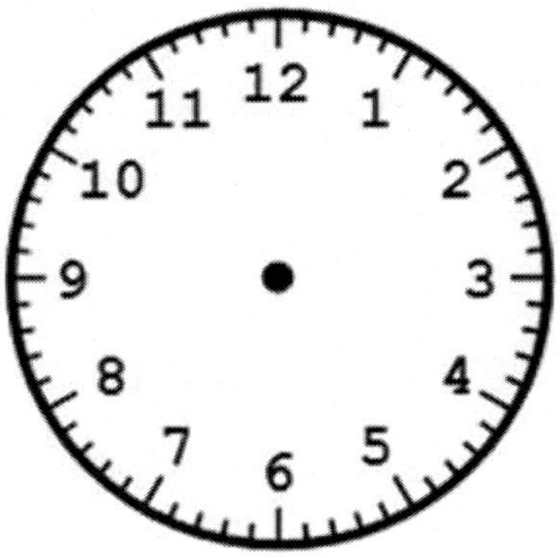

3:45

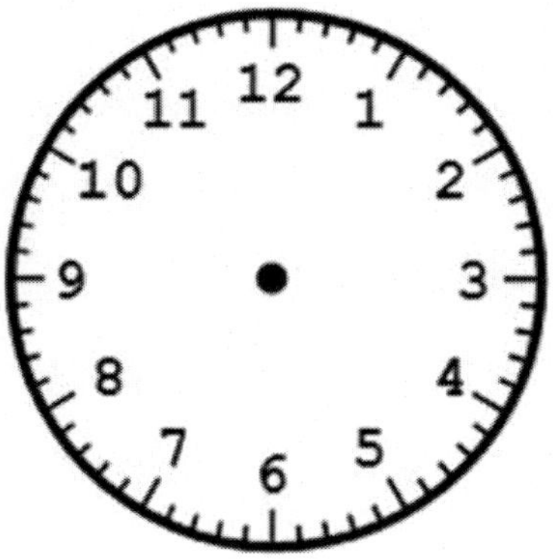

Name: ______________________________

Draw the hands on each clock to show the time.

3:15

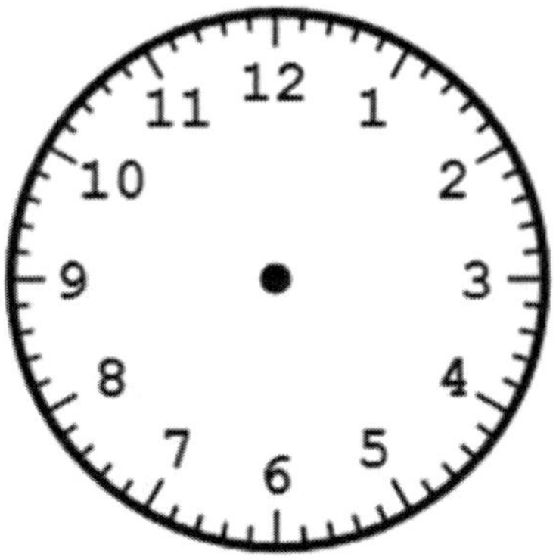

a quarter till 1

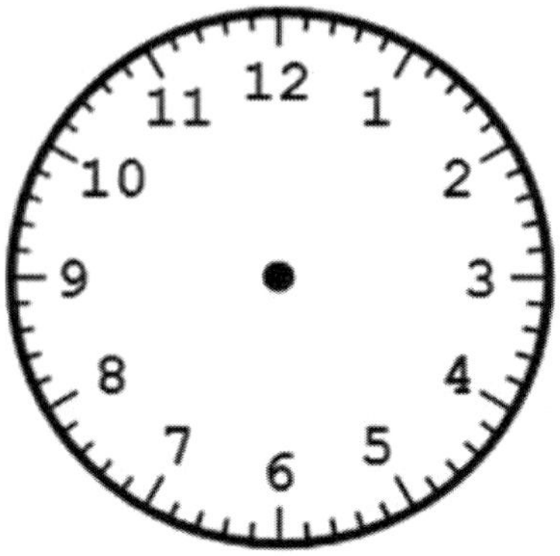

a quarter after 4

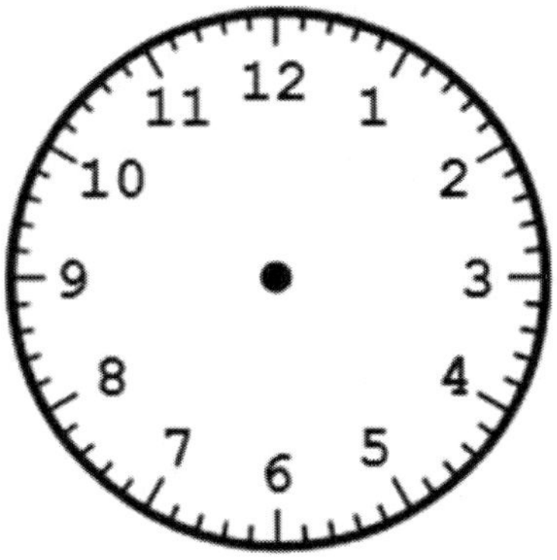

6:45

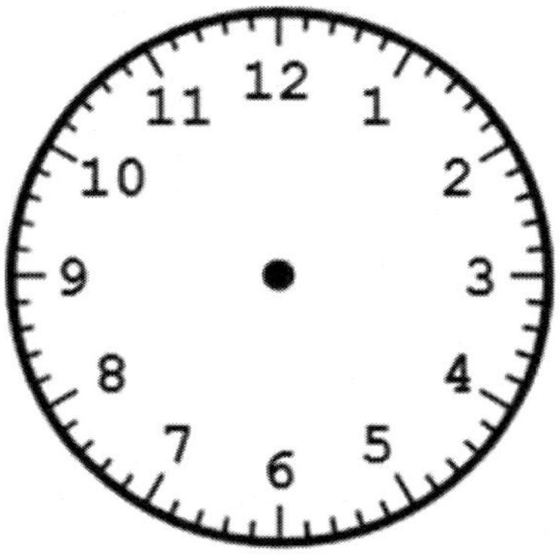

9:45

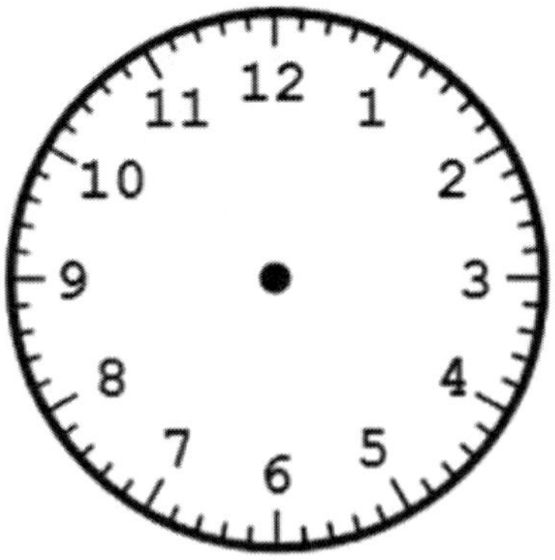

a quarter till 10

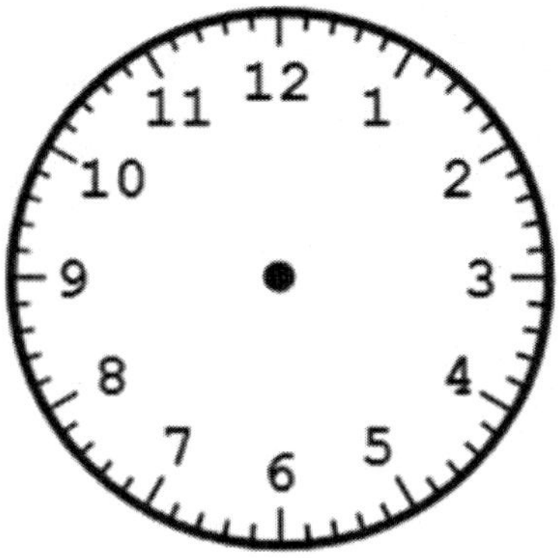

Name: ______________________________

Draw the hands on each clock to show the time.

1:45

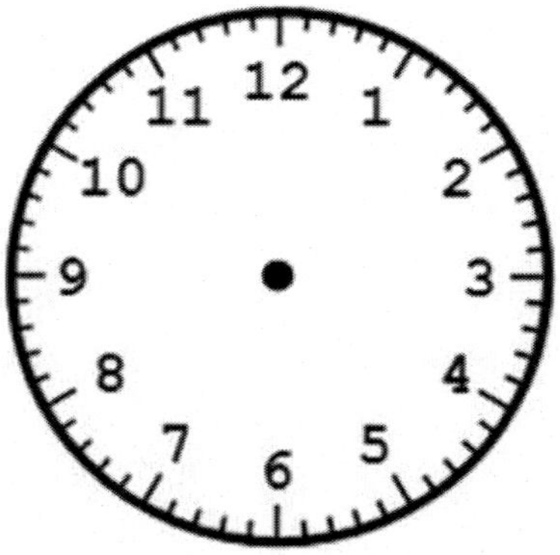

a quarter after 2

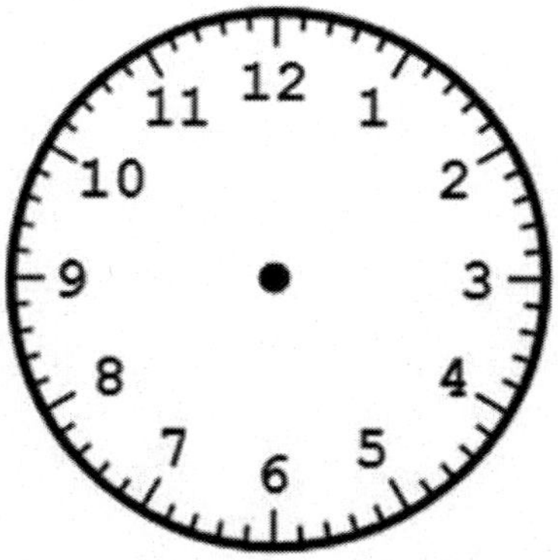

a quarter after 8

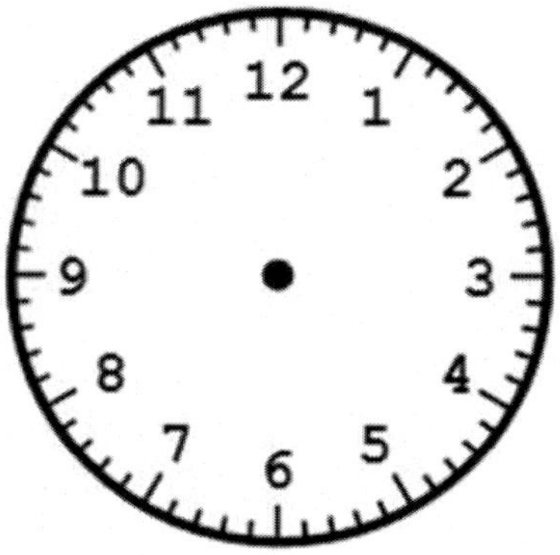

1:45

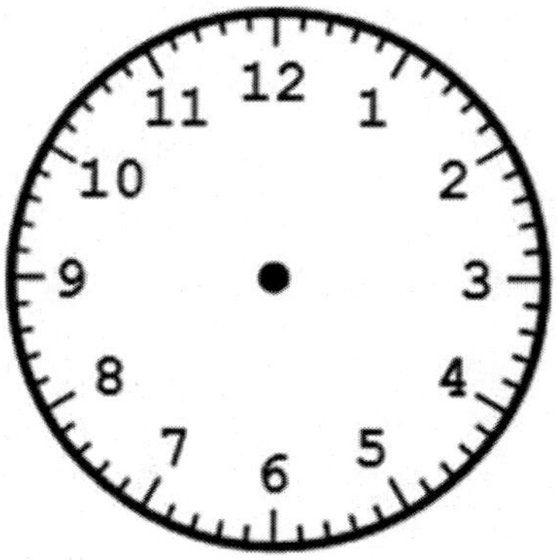

5:45

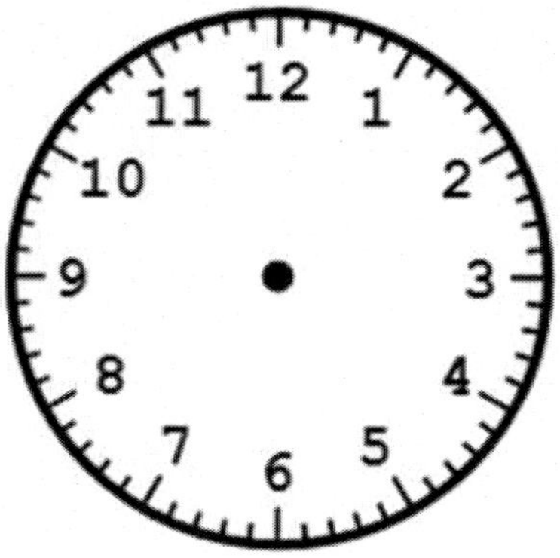

a quarter till 8

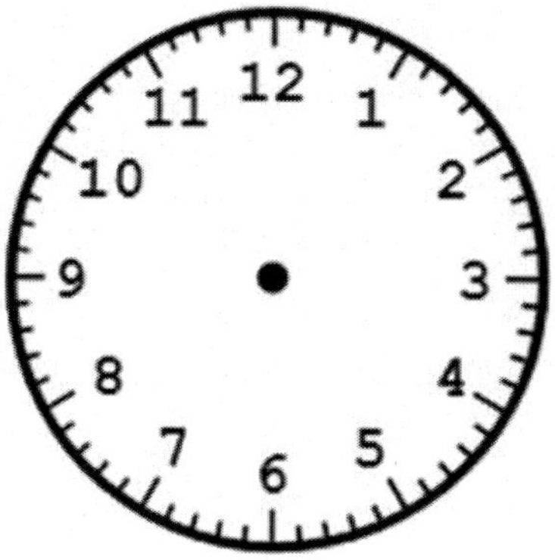

Name: ______________________________

Draw the hands on each clock to show the time.

a quarter after 11

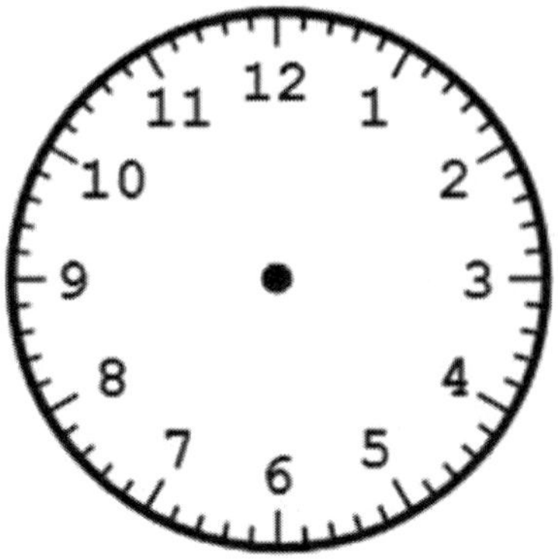

2:45

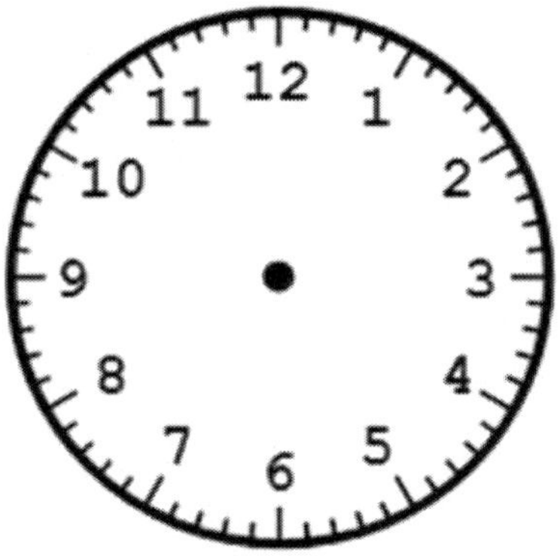

10:45

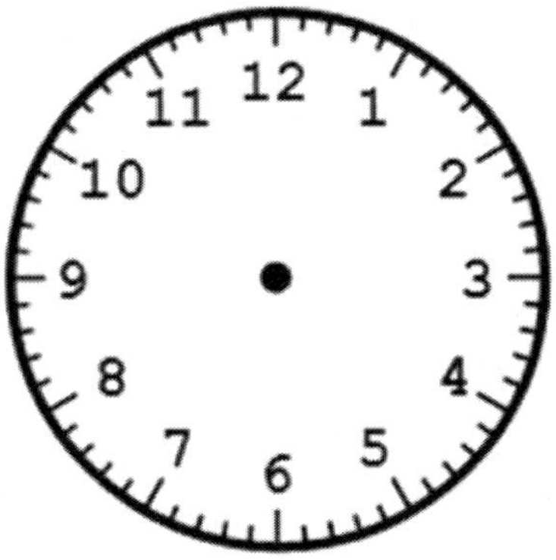

a quarter till 9

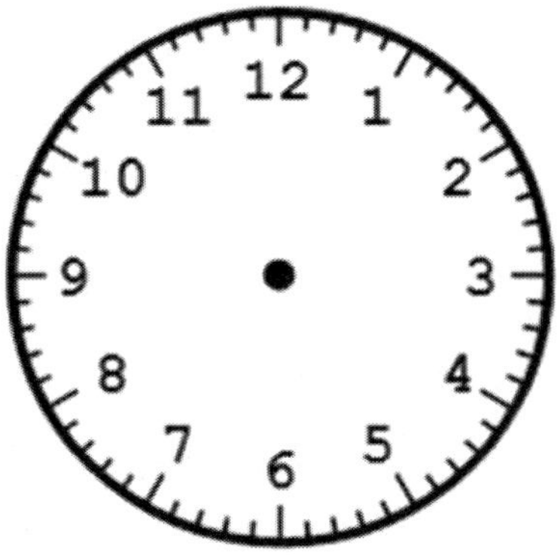

5:15

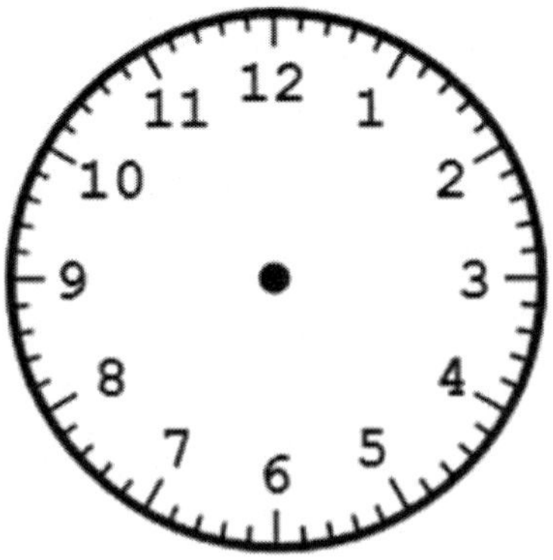

a quarter after 7

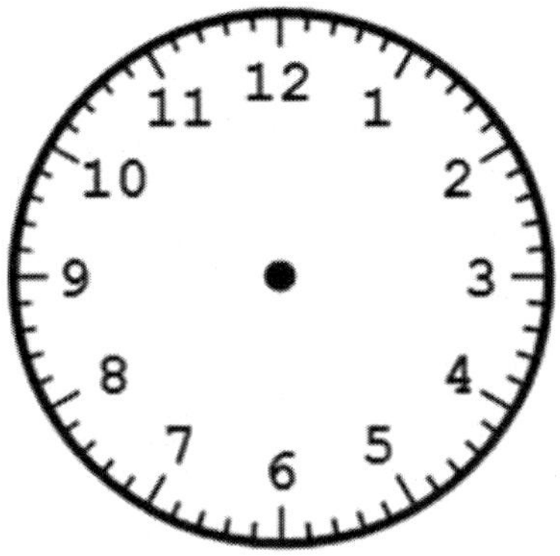

Name: ____________________

Draw the hands on each clock to show the time.

a quarter
till 5

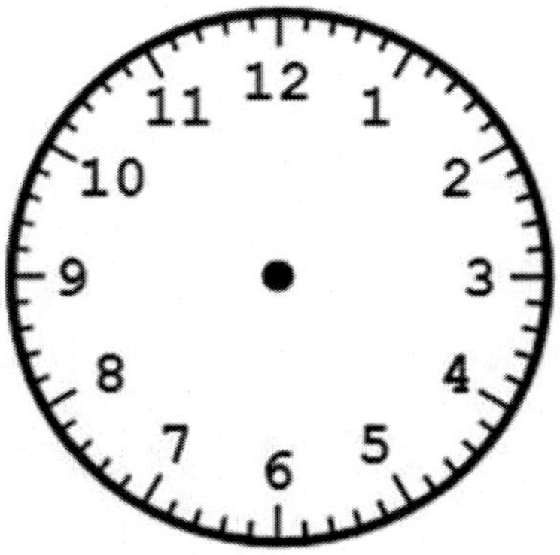

a quarter
after 6

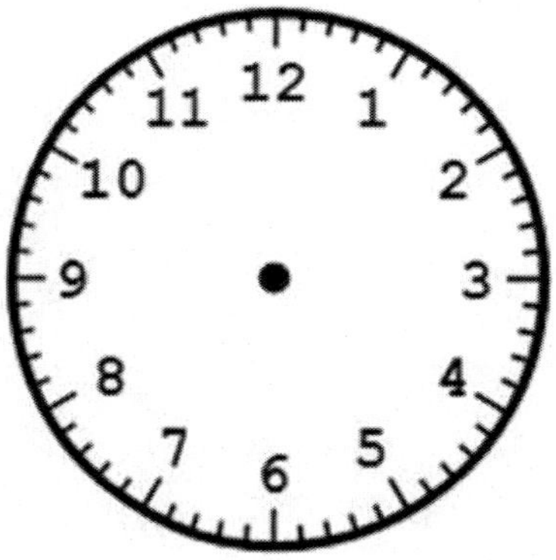

11:45

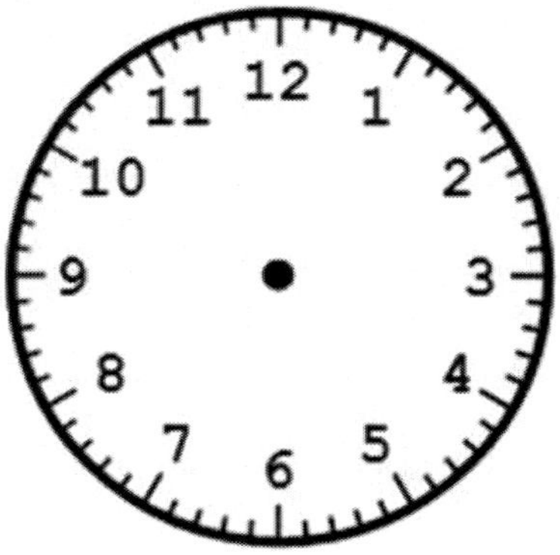

4:45

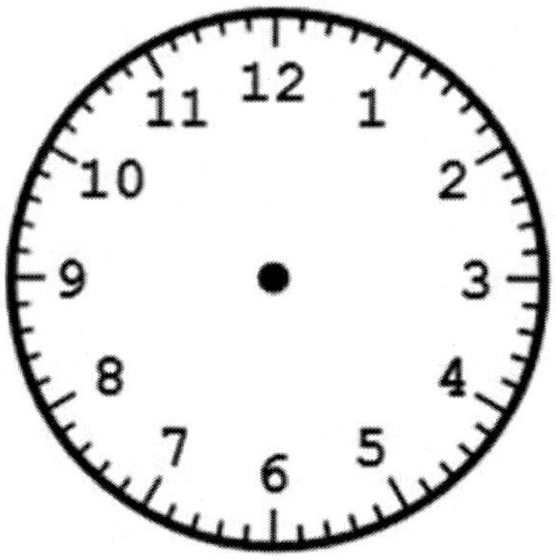

a quarter
after 10

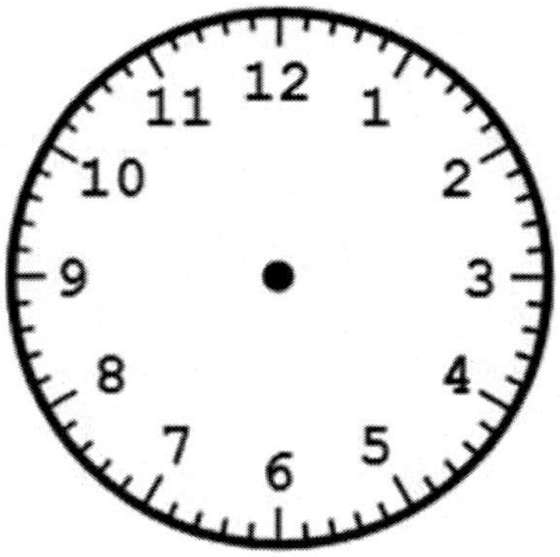

11:15

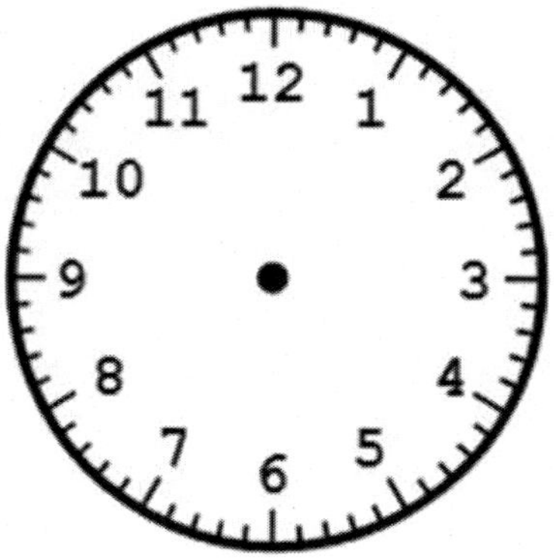

UNIT 12 - MINUTES

The clock is divided into smaller parts so we can tell the exact time.

Every hour has 60 minutes. Tiny lines located between the numbers on the clock face mark minutes.

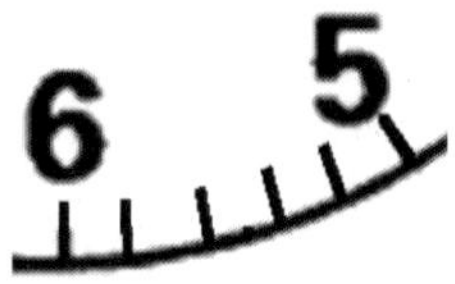

If you count all of the lines on the clock, they would equal 60. There are 60 minutes in 1 hour.

The big hand or minute hand points to one of the line places on the clock every minute.

Sometimes, you will need to know the exact time. To figure out the exact time, skip count by 5 until you reach the last whole number before the big hand.

Count the individual lines to find the exact minute.

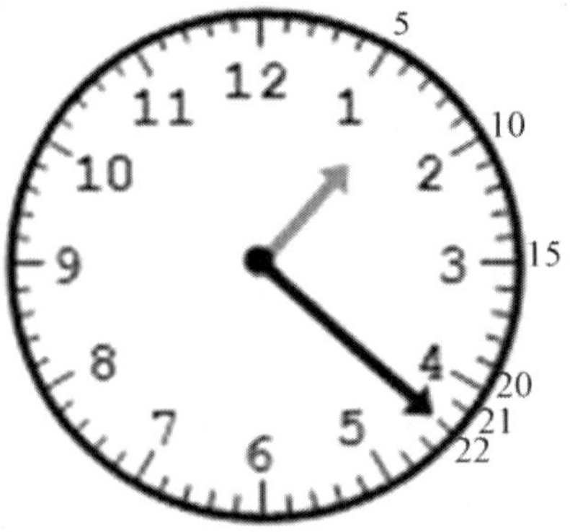

The time shown on the clock is 1:22.

Name: ____

Write the exact time shown by each clock.

1.

______ : ______

2.

______ : ______

3.

______ : ______

4.

______ : ______

5.

______ : ______

6.

______ : ______

Name: ____

Write the exact time shown by each clock.

1.

_______:_______

2.

_______:_______

3.

_______:_______

4.

_______:_______

5.

_______:_______

6.

_______:_______

Name: ____

Write the exact time shown by each clock.

1. ______:______

2. ______:______

3. ______:______

4. ______:______

5. ______:______

6. ______:______

Name: ______________________________

Write the exact time shown by each clock.

1. _____ : _____

2. _____ : _____

3. _____ : _____

4. _____ : _____

5. _____ : _____

6. _____ : _____

Name: ______________________________

Write the exact time shown by each clock.

1.

____ : ____

2.

____ : ____

3.

____ : ____

4.

____ : ____

5.

____ : ____

6.

____ : ____

Name: ______________________________

Write the exact time shown by each clock.

1.

______ : ______

2.

______ : ______

3.

______ : ______

4.

______ : ______

5.

______ : ______

6.

______ : ______

Name: ______________________________

Write the exact time shown by each clock.

1.

_____ : _____

2.

_____ : _____

3.

_____ : _____

4.

_____ : _____

5.

_____ : _____

6.

_____ : _____

Name: ______________________________

Write the exact time shown by each clock.

1. _____:_____

2. _____:_____

3. _____:_____

4. _____:_____

5. _____:_____

6. _____:_____

Name: ______________________________

Draw the hands on the clock to show the time.

5:19

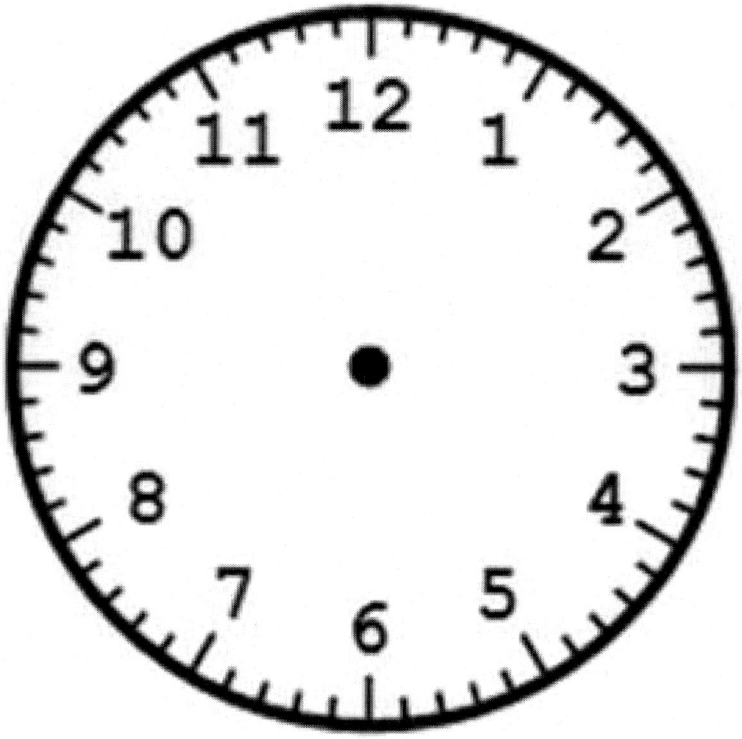

6:29

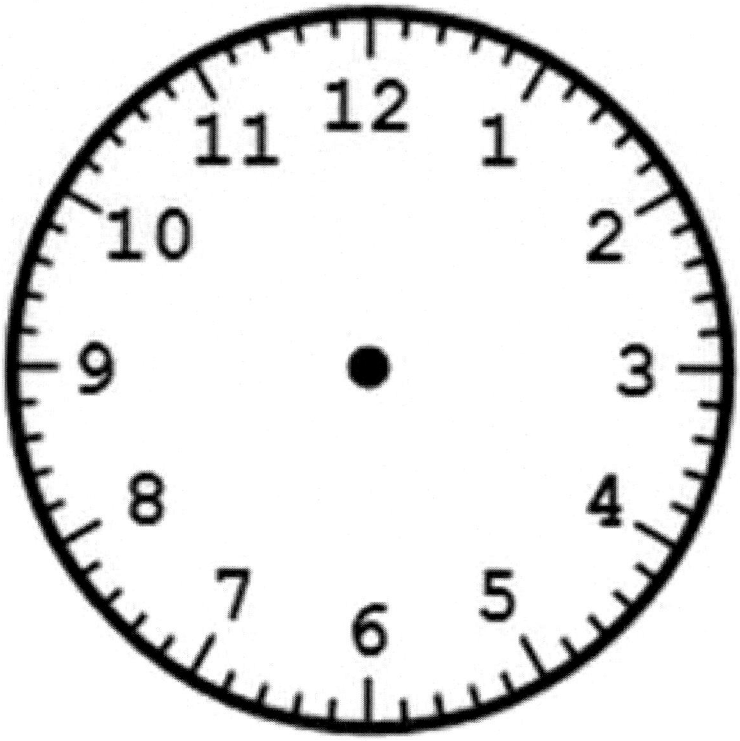

Name: ______________________________

Draw the hands on the clock to show the time.

1:44

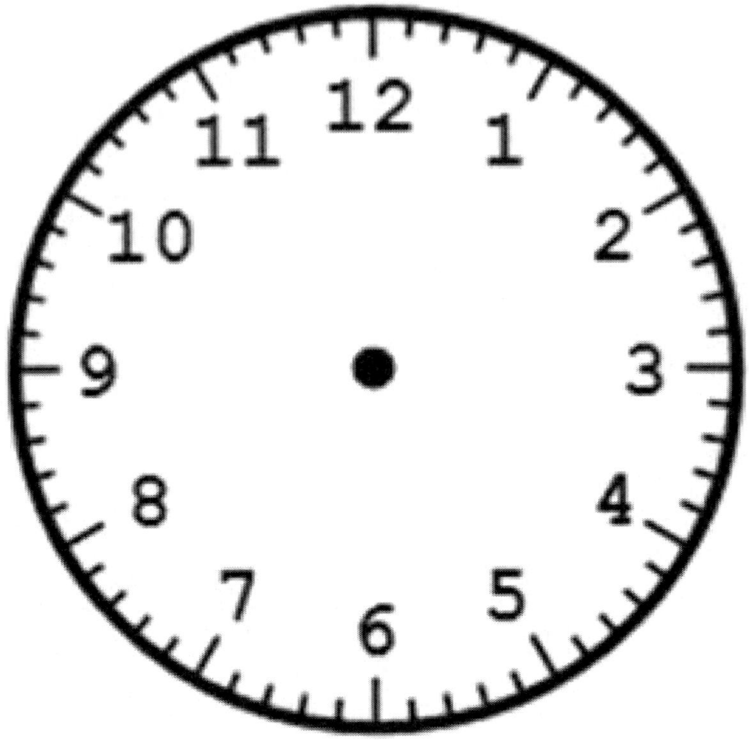

3:09

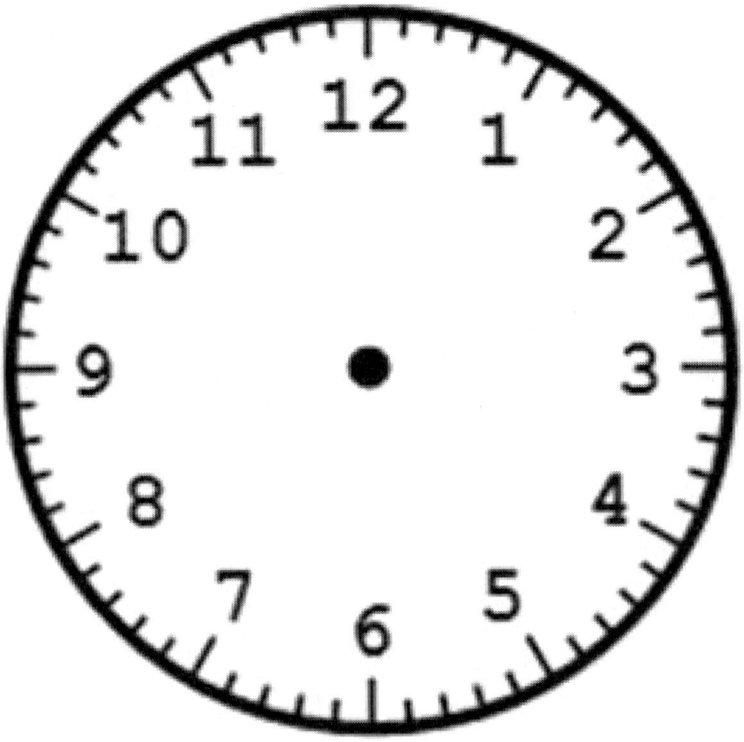

Name: ____

Draw the hands on the clock to show the time.

7:08

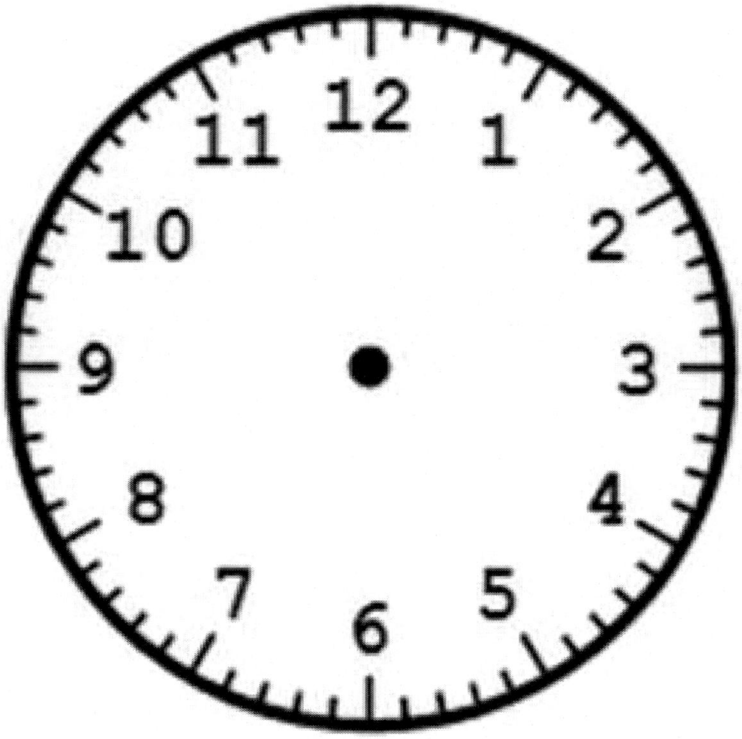

10:37

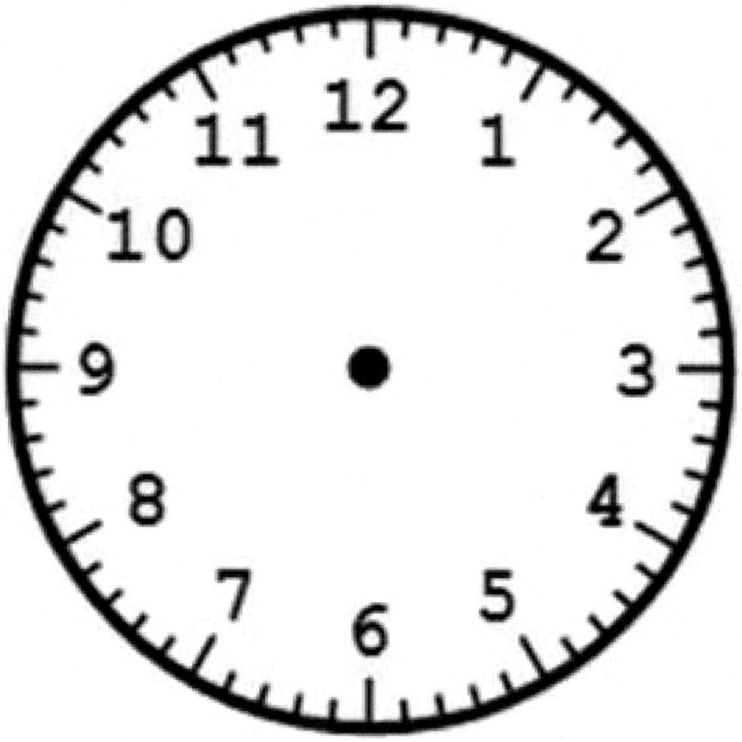

Name: ____________________

Draw the hands on the clock to show the time.

12:46

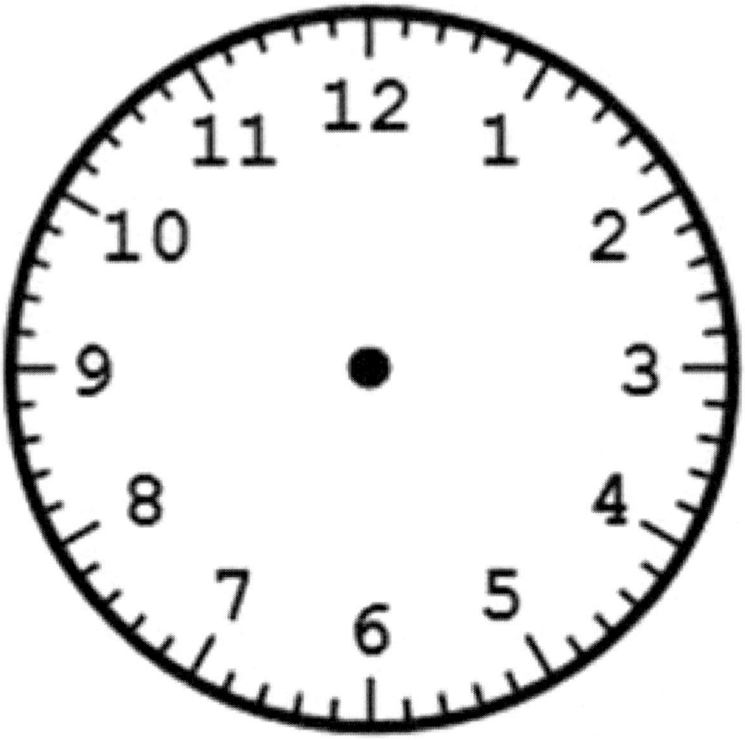

5:31

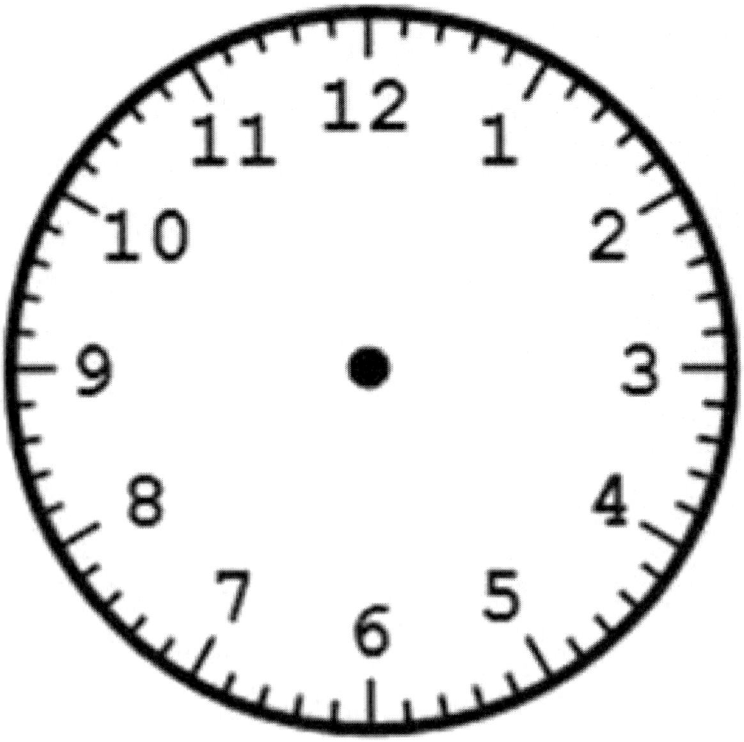

Name: ______________________________

Draw the hands on the clock to show the time.

1:16

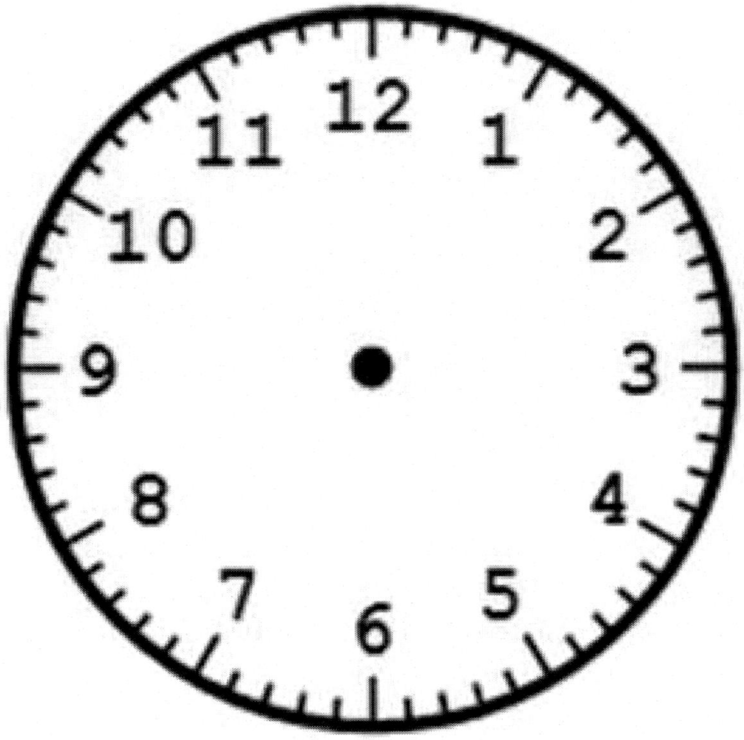

11:03

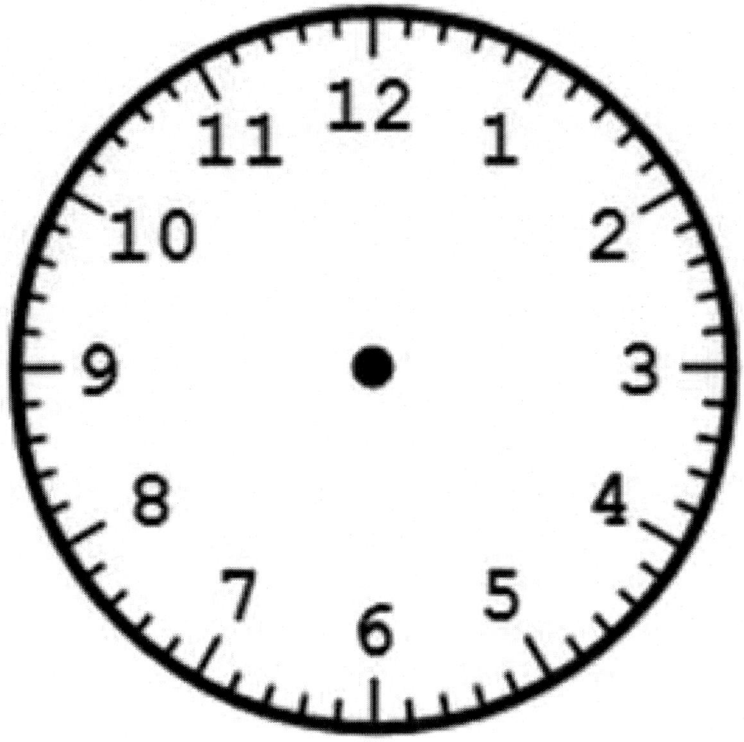

Name: ____________________

Draw the hands on the clock to show the time.

3:42

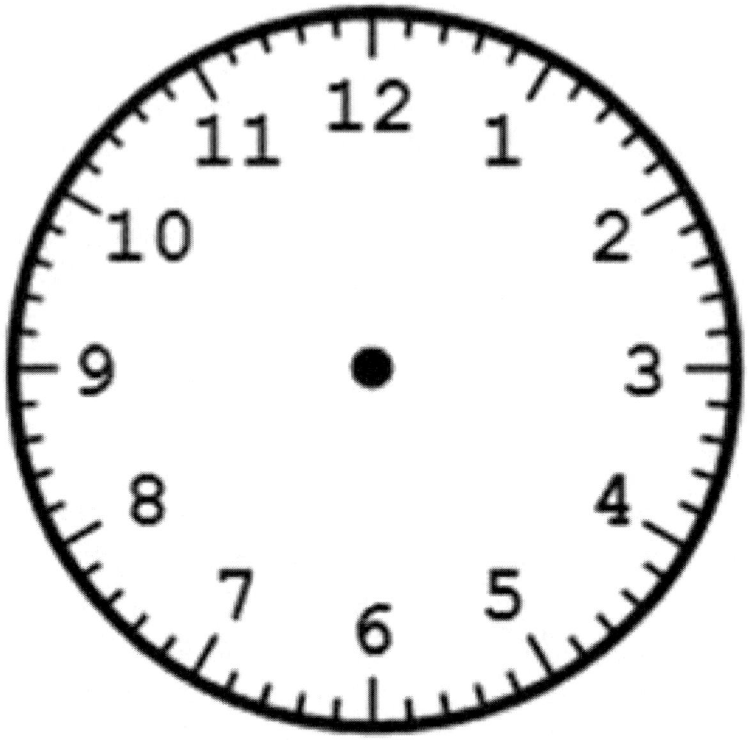

6:38

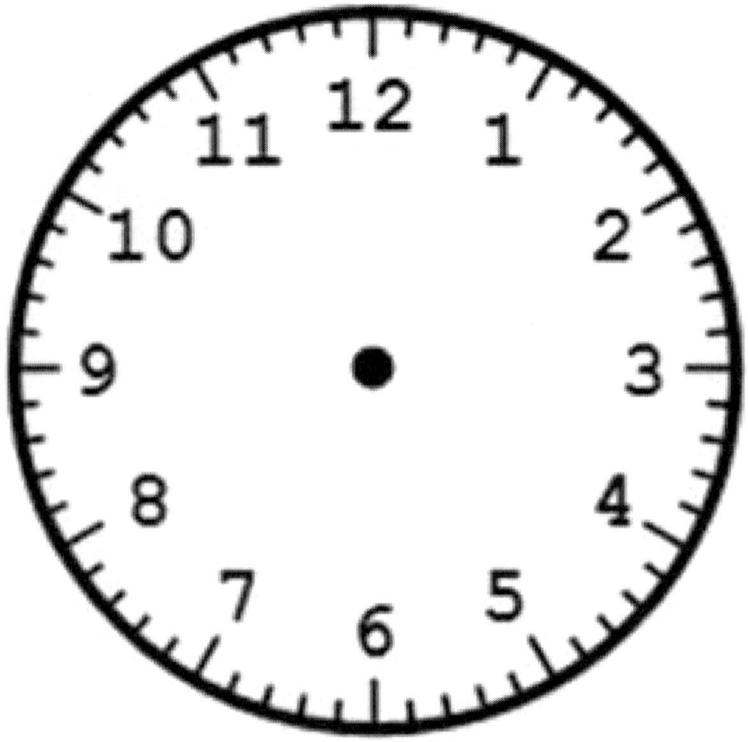

Name: ______________________________

Draw the hands on the clock to show the time.

9:49

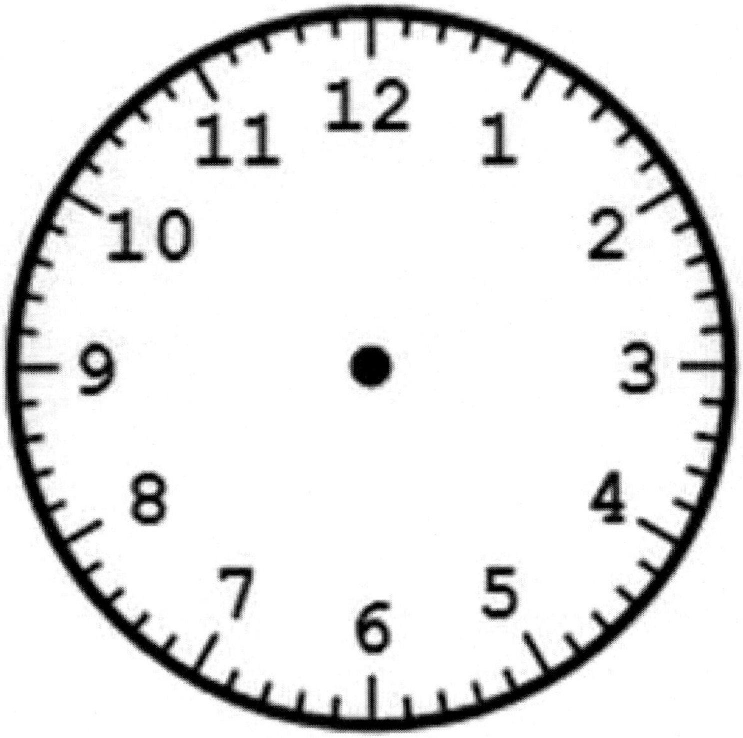

7:51

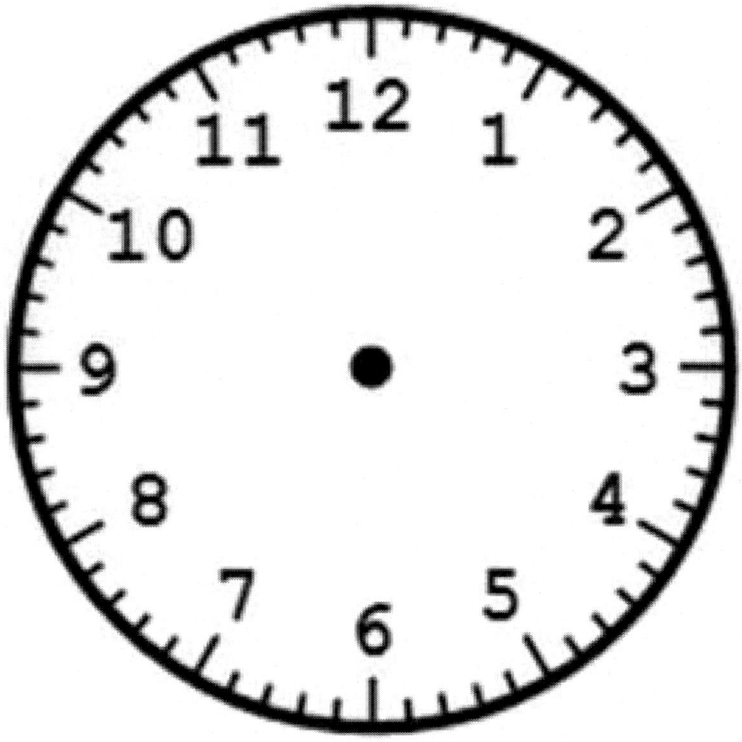

Name: ______________________________

Draw the hands on the clock to show the time.

6:21

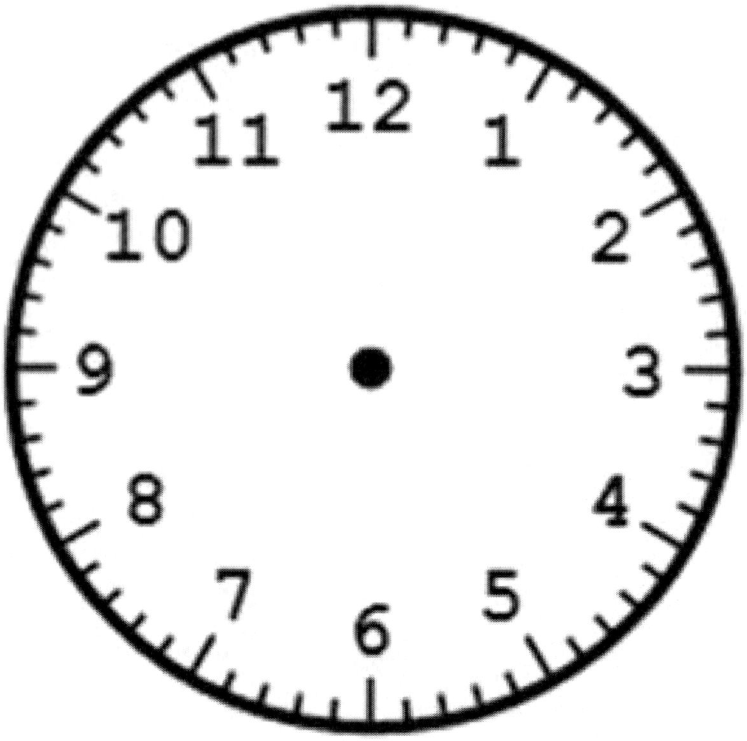

1:04

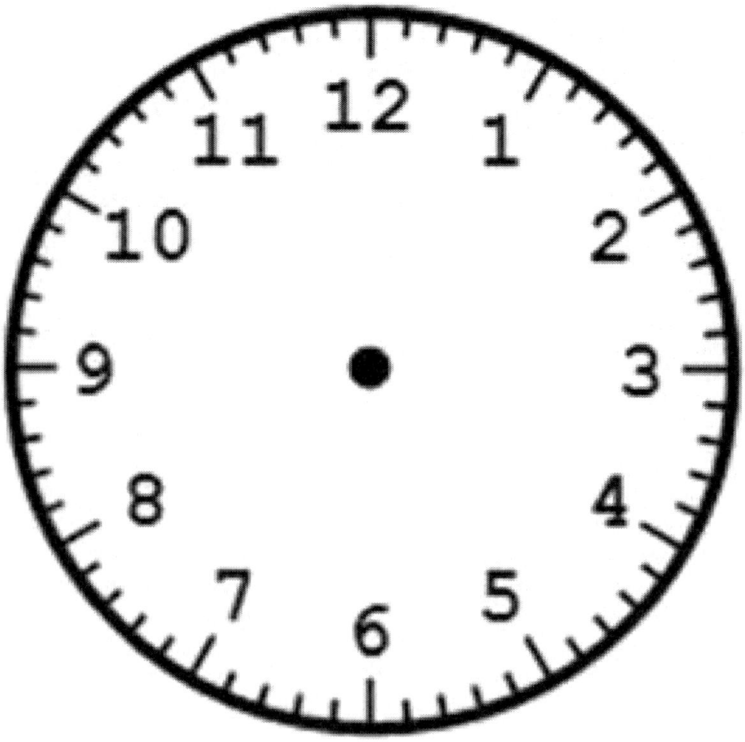

Name: ______________________________

Draw the hands on the clock to show the time.

11:58

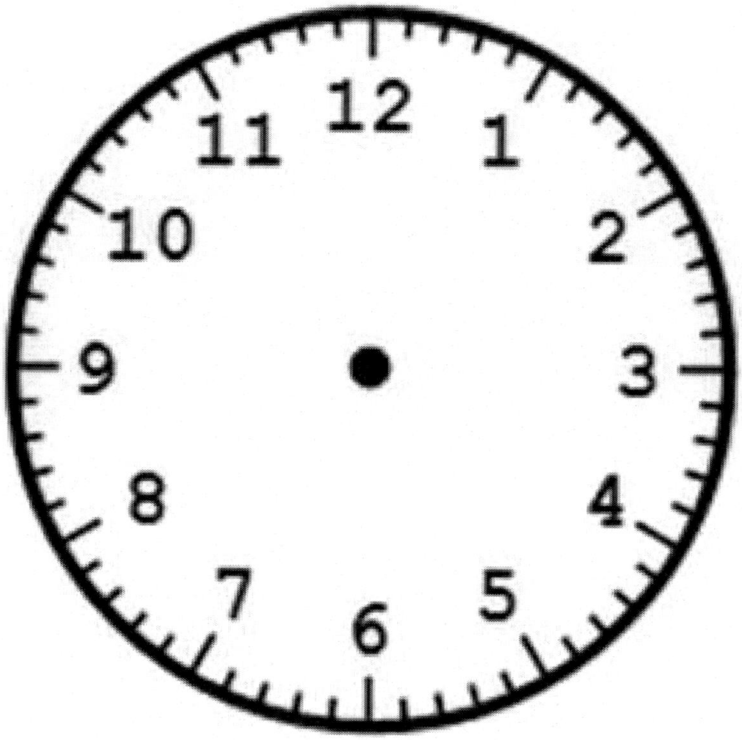

10:47

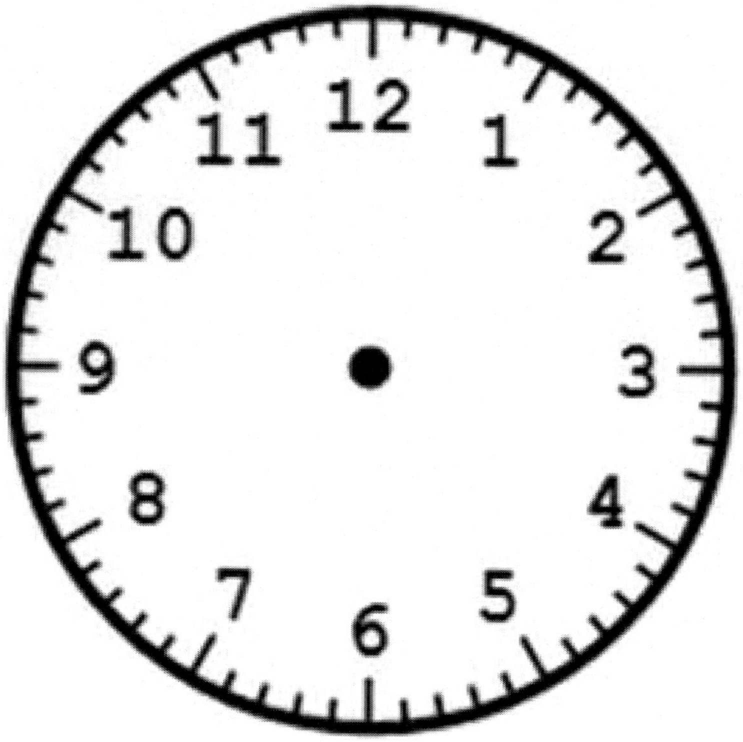

Name: ______________________________

Draw the hands on the clock to show the time.

2:07

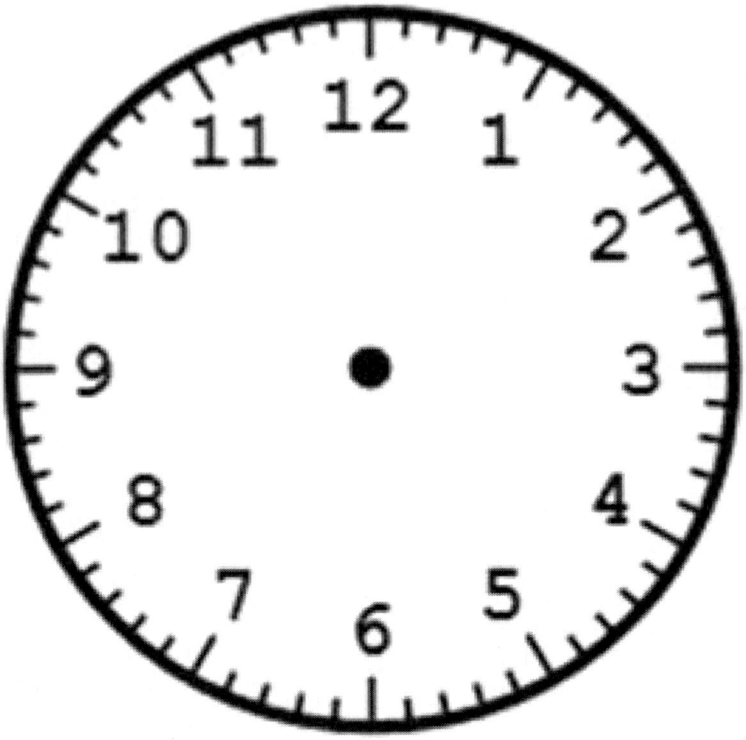

10:13

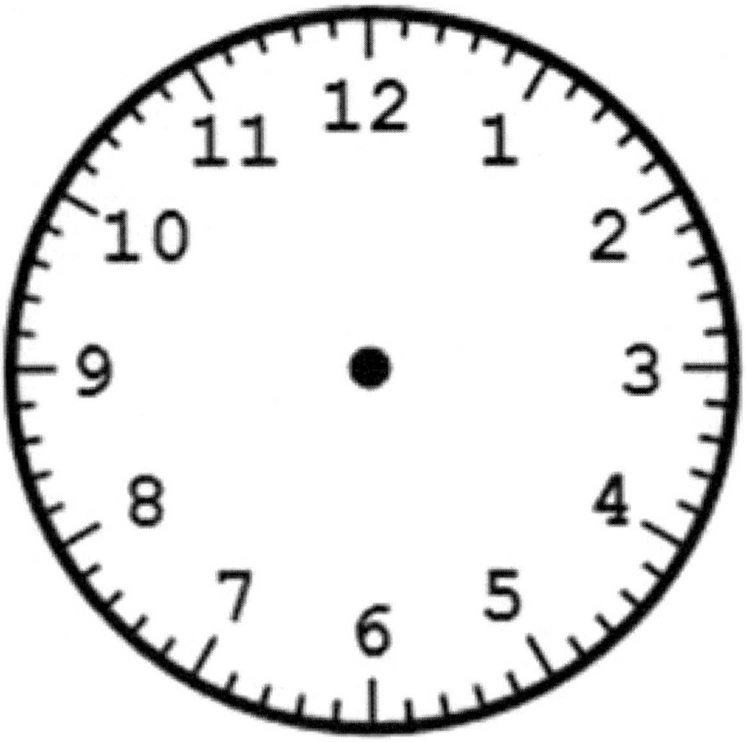

UNIT 13 – AM and PM

There are 24 hours in a day. There are only 12 numbers on a clock face. The hands go the whole way around the clock two times during one day.

Since the little hand points to each hour twice a day, we have to have a way to tell these hours apart.

We call the morning or daytime hours a.m.

We call the evening or night time hours p.m.

8 a.m. is the time of the morning that most people are getting breakfast and preparing to start their day.

8 p.m. is the time that most people are taking their nighttime baths and getting ready to end their day.

We switch from a.m. to p.m. every time both hands reach the 12 on the clock.

12:00 in the daytime is called noon. At 12:00 noon, we switch to p.m.

12:00 at night is called midnight. At 12:00 midnight we switch to a.m.

Name: ____

Answer each question, including AM or PM.

1. I eat lunch at: ______________________________

2. I wake up at: ______________________________

3. Midnight is: ______________________________

4. I start school at: ______________________________

5. I finish school at: ______________________________

6. The sun rises at: ______________________________

7. The sun sets at: ______________________________

Name: ____

Weekdays are Monday, Tuesday, Wednesday, Thursday, and Friday.

Create a schedule for a yourself on a weekday, including AM and PM.

AM Wake Up

Name: ____

The weekend days are Saturday and Sunday.

Create a schedule for a yourself on a weekend day, including AM and PM.

AM Wake Up

Name: ____

Answer each question, including AM or PM.

1. I eat breakfast at: ______________________________

2. My favorite show starts at: ______________________

3. Noon is: _______________________________________

4. I have dinner at: ________________________________

5. I go to bed at: _________________________________

6. My parents go to bed at: _________________________

7. My favorite time of day is: _______________________

UNIT 14 – ELAPSED TIME

Sometimes you will need to figure out how much time has passed, or elapsed, from one thing to another.

Sometimes you will need to figure out how much time is left before an event starts.

Look at the clocks.

Start

End

The clock on the left shows an activity started at 10:45.

The clock on the right shows an activity ended at 12:20

Add one hour to the clock on the left.

10:45 + 1 hour = 11:45

Add another hour to the clock on the left

11:45 + 1 hour = 12:45

12:45 is after the clock on the right so we know we only have to add 1 hour to our start time.

Count the minutes from the 10:45 + 1 hour = 11:45 until you get to 12:20.

11:45 + 5 = 11:50

11:50 + 5 = 11:55

11:55 + 5 = 12:00

12:00 + 5 = 12:05

12:05 + 5 = 12:10

12:10 + 5 = 12:15

12:15 + 5 = 12:20 end time

Put them together.

1 hour + 5 + 5 + 5 + 5 + 5 + 5 + 5 = 1 hour 35 minutes.

The time elapsed from the start to the end is 1 hour and 35 minutes.

Name: ____

Figure out the elapsed time from each start clock on the left to each end clock on the right.

Start

End

HOURS:

MINUTES:

Start

End

HOURS:

MINUTES:

Name: ______________________________

Figure out the elapsed time from each start clock on the left to each end clock on the right.

Start

End

Start

End

Name: ______________________________

Figure out the elapsed time from each start clock on the left to each end clock on the right.

Start

End

Start

End

Name: ______________________________

Figure out the elapsed time from each start clock on the left to each end clock on the right.

Start

End

Start

End

Name: ____

Figure out the elapsed time from each start clock on the left to each end clock on the right.

Start

End

Start

End

Name: ____

Figure out the elapsed time from each start clock on the left to each end clock on the right.

Start

End

Start

End

Name: ______________________________

Figure out the elapsed time from each start clock on the left to each end clock on the right.

Start

End

Start

End

Name: ____

Figure out the elapsed time from each start clock on the left to each end clock on the right.

Start

End

Start

End

Name: ____

Figure out the elapsed time from each start clock on the left to each end clock on the right.

Start

End

Start

End

Name: ____

Figure out the elapsed time from each start clock on the left to each end clock on the right.

Start

End

Start

End

Name: ____

Figure out the elapsed time from each start clock on the left to each end clock on the right.

Start

End

Start

End

Name: ____

Figure out the elapsed time from each start clock on the left to each end clock on the right.

Start

End

Start

End

Name: ____

Figure out the elapsed time from each start clock on the left to each end clock on the right.

Start

End

Start

End

Name: ____

Figure out the elapsed time from each start clock on the left to each end clock on the right.

Start

End

Start

End

Name: ____

Figure out the elapsed time from each start clock on the left to each end clock on the right.

Start

End

Start

End

CALULATING ELAPSED TIME

Sometimes you will not know the end time. You will know what time something starts and how long it lasts. You will need to figure out the end time on your own.

Pretend you are going to a movie with your friend. Your mother is going to pick you up when the movie ends.

You know the movie starts at 1:15.

You know the movie lasts 1 hour and 35 minutes.

You need to figure out what time the movie ends so you can tell your mother when she should pick you up.

Write the start time:	1:15
Add the minutes:	+ :35
	1:50
Add the hours:	+ 1:00
	2:50

The movie will end at 2:50.

Sometimes when you add the minutes, they will equal more than a full hour or more than 60. When that happens, you must carry the minutes to the hour column just like in regular math except you carry when you reach 60 instead of 100.

You know the movie starts at 1:15.

You know the movie lasts 1 hour and 55 minutes.

You need to figure out what time the movie ends so you can tell your mother when she should pick you up.

Write the start time:	1:15
Add the minutes:	+ :55
	1:70

1:70 is not a time.

Convert 70 minutes to hours and minutes	70
	- 60 (1 hour)
	10 (minutes)

Write the start time:	1:15
Add the carried hour:	+ 1:00
	2:15
Add the remainder minutes:	+ :10
The movie will end at:	2:25
Add the original hour:	+ 1:00
The movie will end at:	**3:25**

Name: ____

Figure out the elapsed time. Use the blank clocks to help you calculate the math.

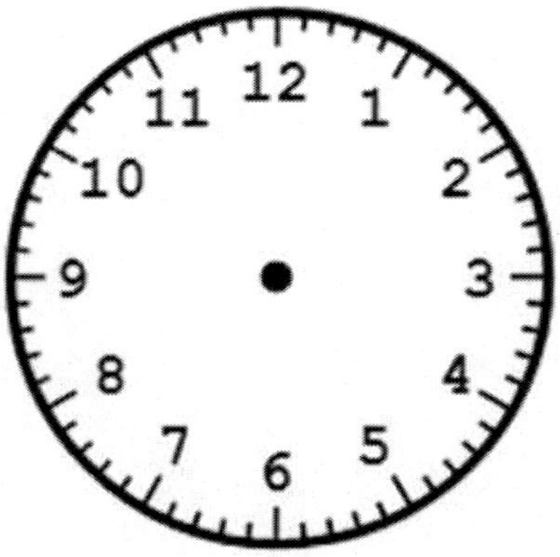

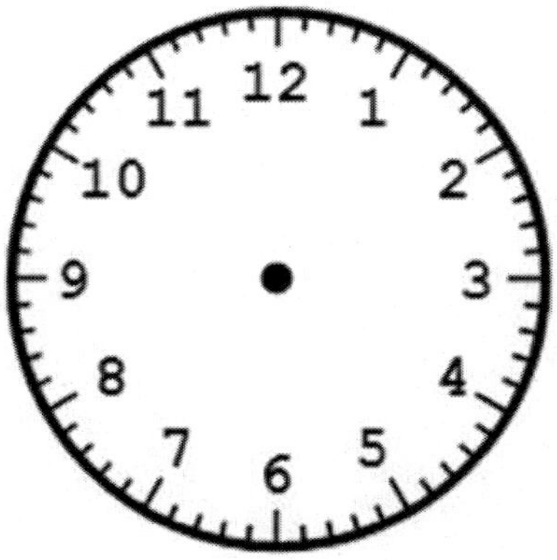

Start End

Karate class starts at 5:00 and lasts 1 hour and 30 minutes. What time does karate class end?

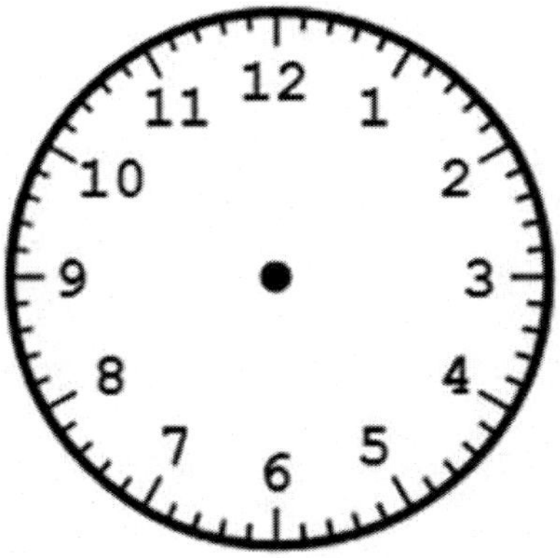

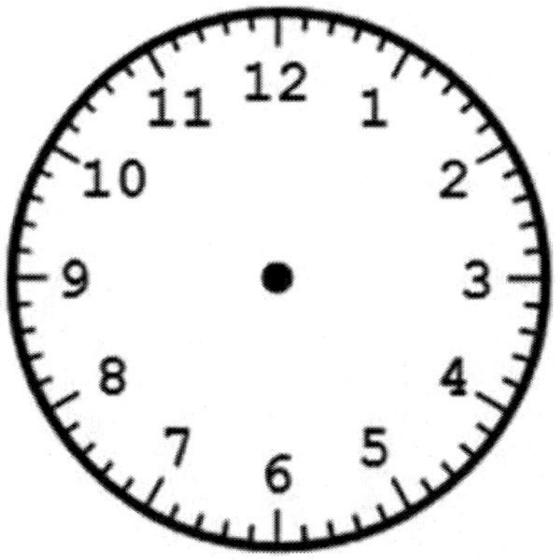

Start End

Sarah leaves for the store at 1:20. It takes 15 minutes to drive to the store. What time will Sarah arrive at the store?

Name: ______________________________

Figure out the elapsed time. Use the blank clocks to help you calculate the math.

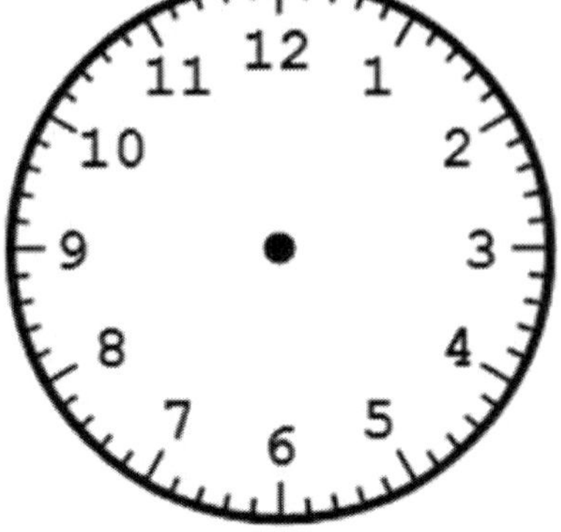

Start End

The movie begins at 1:50 and lasts 1 hour and 25 minutes. What time does the movie end?

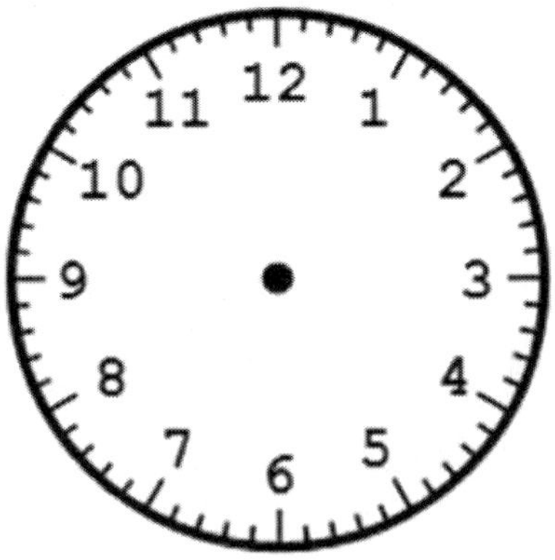

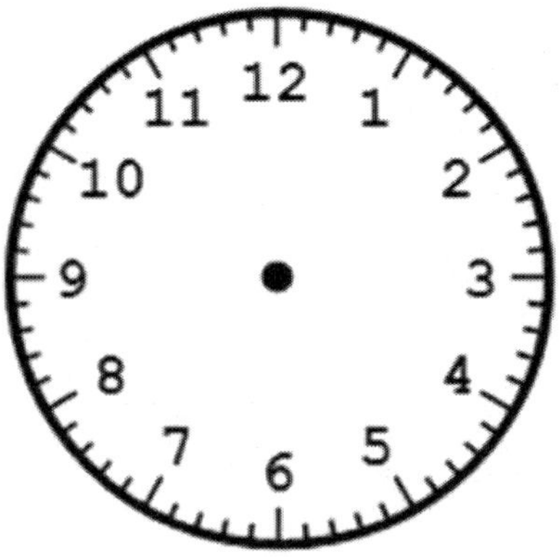

Start End

Ben's class starts at 6:00 and lasts for 1 hour and 45 minutes. What time will Ben's class be finished?

Name: ____

Figure out the elapsed time. Use the blank clocks to help you calculate the math.

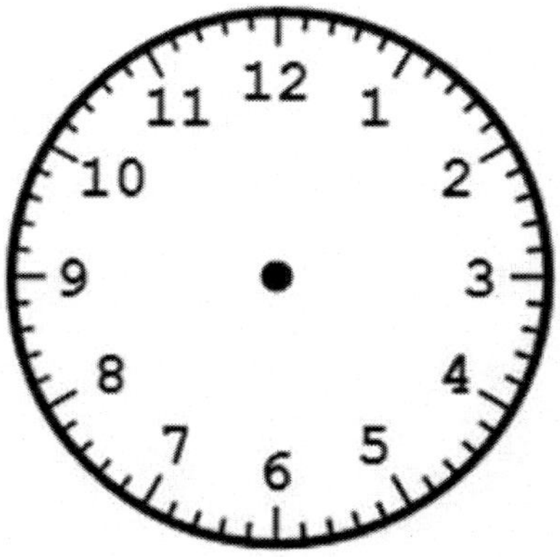

Start

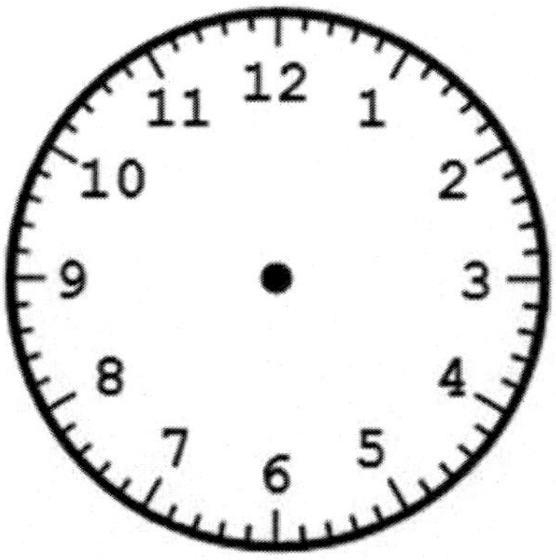

End

Our field trip begins at 9:00 a.m. and lasts 3 hours. What time does the field trip end?

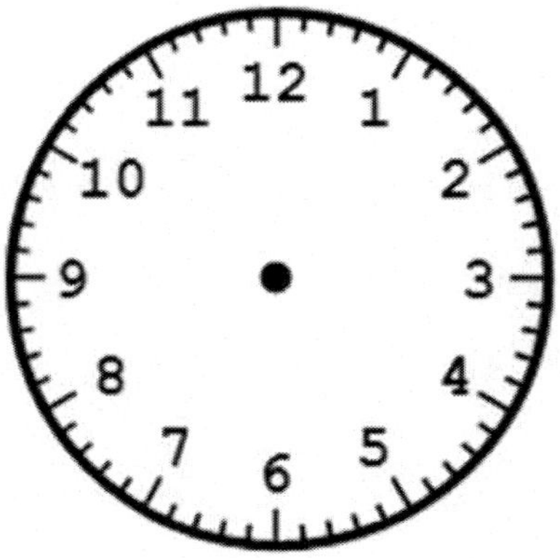

Start

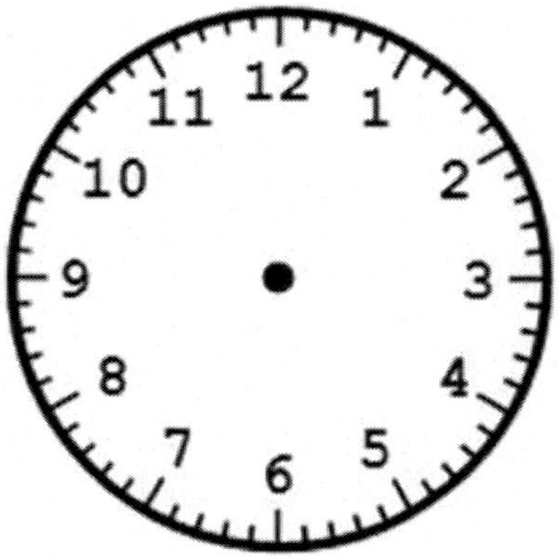

End

Erica left for the park at 3:30. Her mom told her she could play for 2 hours. What time should Erica go home?

Name: ____

Figure out the elapsed time. Use the blank clocks to help you calculate the math.

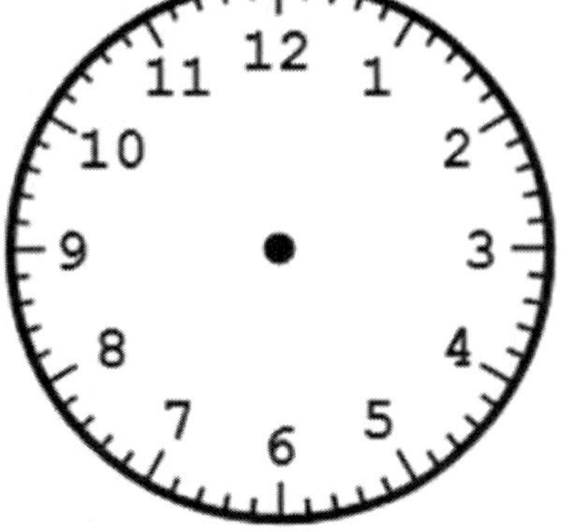

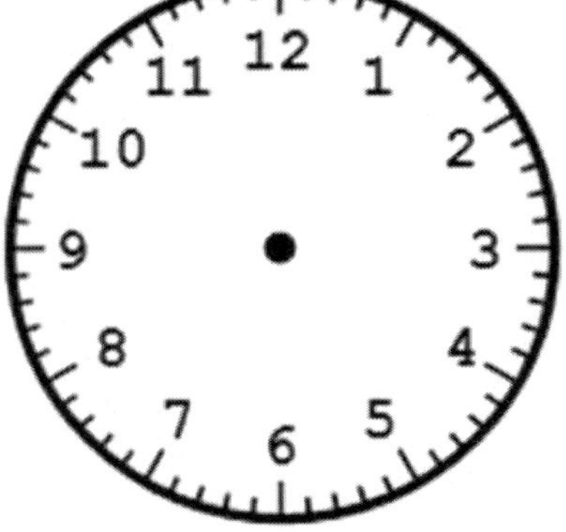

Start End

Math class begins at 2:05 and lasts 55 minutes. What time does math class end?

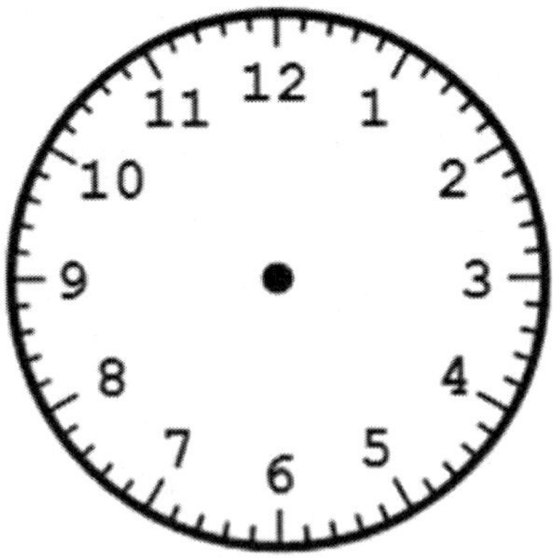

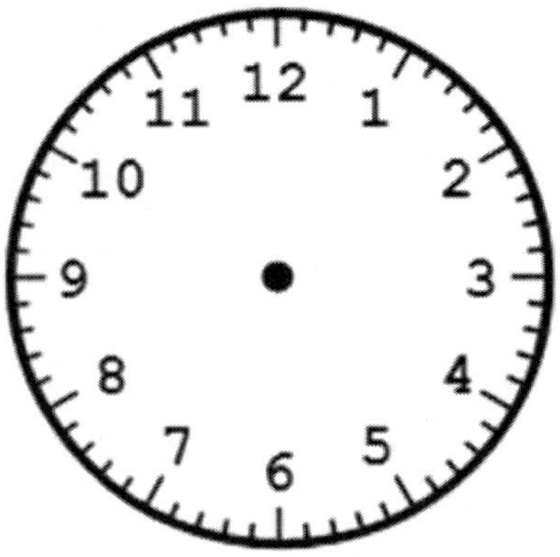

Start End

The teacher handed out the test at 1:15. The rules say the students must complete the test within 60 minutes. What time will the test end?

Name: ____

Figure out the elapsed time. Use the blank clocks to help you calculate the math.

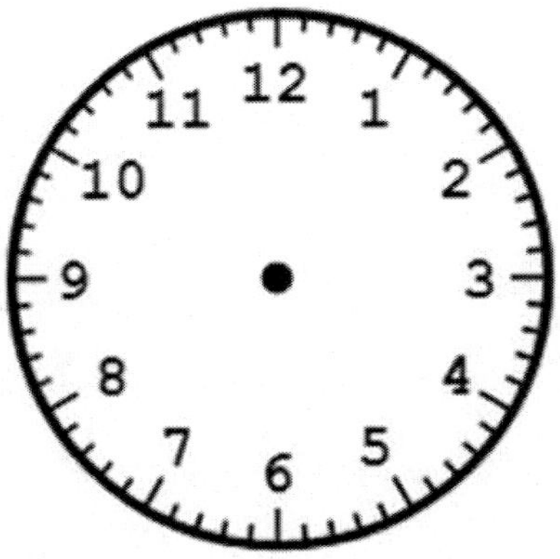

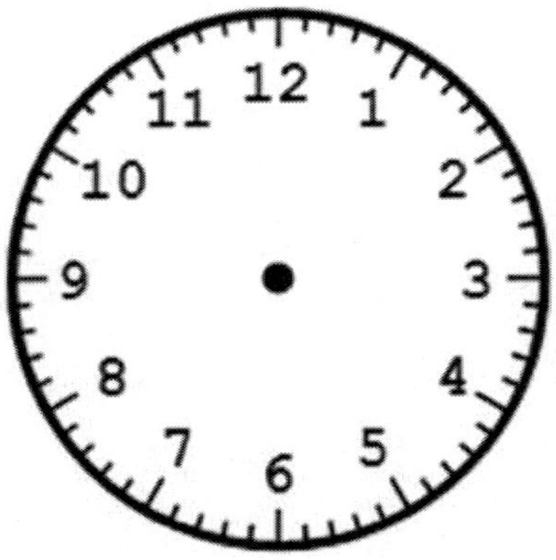

Start End

It is 3:15. The flight leaves in 2 and 1/2 hours. What time does the flight leave?

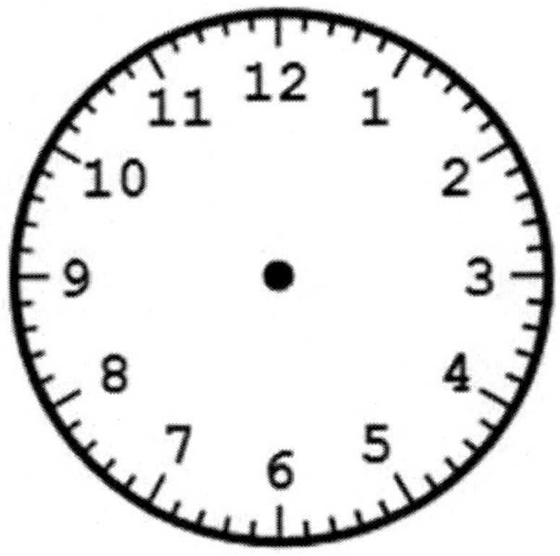

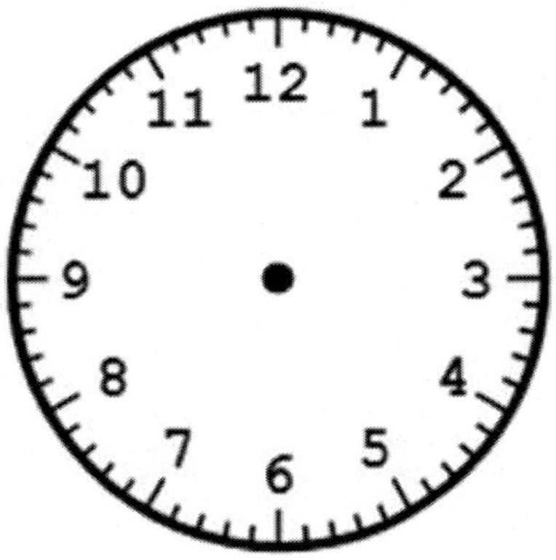

Start End

Tim gets home from school at 3:35. It takes him 1 hour to complete his homework and 15 minutes to feed his pets. When he is done with his work, he can play. What time will Tim be finished with his work so he can play?

Name: ______________________________

Figure out the elapsed time.

Lisa went on a bike ride at 1:40. She rode her bike for 3 hours and 10 minutes. What time did Amanda finish riding her bike?

Sarah walked to the store. She left the house at 5:10. She arrived back home at 5:55. How long did it take Amanda to walk to the store and back home?

Name: ____

Figure out the elapsed time.

The television show begins at 7:00. The show ends at 8:30. How long did the show last?

Class starts at 8:10 AM. Lunch begins 3 hours and 15 minutes later. What time is lunch?

Name: ____

Figure out the elapsed time.

It is 7:15. John is coming over in 1 hour and 20 minutes. What time will John arrive?

It is 6:45. We are going leaving for the zoo at 8:30. How long until it is time to leave for the zoo?

Name: ____

Figure out the elapsed time.

Chris started playing his game at 6:35. He is allowed to play for 2 hours. What time should Chris finish his game?

The test will begin at 1:10. The student has 120 minutes to finish the test. What time will the test end?

Name: ____

Figure out the elapsed time.

Beverly and Caleb started working on their project at 4:18. They finished their project at 6:00. How long did they work on the project?

The speaker started the speech at 12:17. He spoke for 45 minutes. What time did the speaker finish his speech?

Name: ____

Figure out the elapsed time.

The party starts at 6:00. It will take 25 minutes to drive to the party. What time should we leave for the party?

We put the brownies in the oven at 4:39. The brownies need to cook for 55 minutes. What time should we take the brownies out of the oven?

Name: ______________________________

Figure out the elapsed time.

The plane leaves at 12:17. We must arrive at the airport 1 hour and 30 minutes before the plane leaves. What time should we arrive at the airport?

I need to be at the mall by 1:00 PM. The taxi takes 17 minutes to drive from my house to the mall. What time should the taxi leave my house?

ANSWERS

Page 7
1. March
2. November
3. February
4. July
5. May
6. September
7. June
8. January
9. August
10. December
11. October
12. April

Page 8
1. 12
2. October
3. March
4. May
5. January

Page 9
1. December
2. March
3. September
4. June
5. February

Page 10
1. 9th
2. 6th
3. 4th
4. 12th
5. 8th

Page 11
1. 3rd
2. 2nd
3. 7th
4. 5th
5. 1st

Page 15
1. Monday
2. Saturday
3. Sunday
4. Wednesday
5. Tuesday
6. Friday
7. Thursday
8. Monday
9. Thursday
10. Wednesday

Page 16
1. 7
2. Sunday
3. Saturday
4. Wednesday
5. Tuesday
6. Sunday
7. Friday
8. Tuesday
9. Friday
10. Thursday
11. Monday
12. Saturday
13. Wednesday
14. Monday

Page 17
1. fourth
2. seventh
3. second
4. third
5. fifth
6. first
7. sixth

Page 20
1. September
2. April
3. November
4. 31
5. February
6. 28
7. 29

Page 21
1. 31
2. 28
3. 31
4. 30
5. 31
6. 30
7. 31
8. 31
9. 30
10. 31
11. 30
12. 31

Page 22
1. 7
2. 31
3. 31
4. 31
5. 31

Page 23
1. 31
2. 28
3. 4
4. 30
5. 30

Page 24
1. 31
2. 31
3. 1
4. 30
5. 30

Page 25
1. Wednesday
2. 5
3. 17th
4. 31
5. Wednesday
6. 4
7. 6th
8. Thursday

Page 26
1. Wednesday
2. 4
3. 8th
4. 31
5. Thursday
6. 4
7. 28th
8. Wednesday

Page 27
1. Tuesday
2. 4
3. 6th
4. 28
5. Friday
6. 4
7. 12th
8. Wednesday

Page 28
1. Thursday
2. 4
3. 14th
4. 30
5. Friday
6. 4
7. 18th
8. Thursday

Page 29 - 59

Answers will vary

Page 62
1. 01-10-2012
2. 11-04-2021
3. 12-29-1999
4. 06-24-2071
5. 04-13-2020
6. 10-30-2011
7. 02-19-2002
8. 08-23-2006
9. 03-03-2033
10. 05-11-1902
11. 07-04-2017
12. 09-28-1954

Page 63
1. November 14, 2017
2. March 27, 2006
3. December 23, 2021
4. January 04, 1999
5. February 19, 2002
6. April 14, 1970
7. August 23, 2006
8. May 31, 2071
9. July 11, 1965
10. June 03, 2070
11. September 23, 1952
12. October 12, 2015

Page 66
1. Jan.
2. Feb.
3. Mar.
4. Apr.
5. May
6. June
7. July
8. Aug.
9. Sept.
10. Oct.
11. Nov.
12. Dec.

Page 67
1. Jan. 01
2. Varied
3. March 17
4. Varied
5. Varied
6. Oct. 31
7. Varied
8. Dec. 31
9. Varied
10. Varied

Page 68
1. Varied
2. Apr. 1
3. Varied
4. Varied
5. Varied
6. Dec. 24
7. Varied
8. Varied
9. Varied
10. Varied

Page 70
1. Sun.
2. Mon.
3. Tues.
4. Wed.
5. Thurs.
6. Fri.
7. Sat.

Page 71
1. Sunday
2. Monday
3. Tuesday
4. Wednesday
5. Thursday
6. Friday
7. Saturday

Page 72
1. 7
2. April
3. New Year's
4. July
5. Tuesday

Page 73
1. 12
2. 366
3. July
4. Thursday
5. 365

Page 76
1. 3:00
2. 1:00
3. 7:00
4. 11:00
5. 10:00
6. 4:00

Page 77
1. 6:00
2. 8:00
3. 12:00
4. 2:00
5. 9:00
6. 5:00

Page 86
1. 12:20
2. 12:45
3. 7:35
4. 4:10
5. 11:50
6. 9:35

Page 87
1. 11:25
2. 2:45
3. 6:35
4. 12:05
5. 3:55
6. 1:15

Page 88
1. 11:40
2. 10:35
3. 7:20
4. 2:15
5. 3:50
6. 3:05

Page 89
1. 7:15
2. 8:55
3. 3:10
4. 2:35
5. 4:20
6. 12:35

Page 90
1. 10:55
2. 1:35
3. 12:15
4. 7:45
5. 6:05
6. 7:10

Page 91
1. 10:45
2. 8:50
3. 9:20
4. 11:35
5. 5:45
6. 6:25

Page 92
1. 7:05
2. 4:35
3. 10:50
4. 2:10
5. 12:40
6. 9:45

Page 103
1. 1:30
   Half Past 1
2. 5:30
   Half Past 5
3. 12:30
   Half Past 12
4. 11:00
   11 o'clock
5. 10:00
   10 o'clock
6. 4:30
   Half Past 4

Page 104
1. 7:30
   Half Past 7
2. 2:00
   2 o'clock
3. 12:00
   12 o'clock
4. 9:30
   Half Past 9
5. 8:30
   Half Past 8
6. 6:30
   Half Past 6

Page 105
1. 8:00
   8 o'clock
2. 11:30
   Half Past 11
3. 10:30
   Half Past 10
4. 7:00
   7 o'clock
5. 3:30
   Half Past 3
6. 2:30
   Half Past 2

Page 119
1. 3:15
   Quarter After 3
2. 12:45
   Quarter Til 1
3. 6:45
   Quarter Til 7
4. 11:15
   Quarter After 11
5. 4:15
   Quarter After 4
6. 4:45
   Quarter Til 5

Page 120
1. 1:45
   Quarter Til 2
2. 2:45
   Quarter Til 3
3. 12:15
   Quarter After 12
4. 7:45
   Quarter Til 8
5. 8:30
   Half Past 8
6. 1:15
   Quarter After 1

Page 121
1. 10:45
   Quarter Til 11
2. 5:15
   Quarter After 5
3. 10:15
   Quarter After 10
4. 2:15
   Quarter After 2
5. 5:45
   Quarter Til 6
6. 11:45
   Quarter Til 12

Page 122
1. 7:15
   Quarter After 7
2. 3:45
   Quarter Til 4
3. 8:45
   Quarter Til 9
4. 6:15
   Quarter After 6
5. 8:15
   Quarter After 8
6. 9:45
   Quarter Til 10

Page 129
1. 6:26
2. 1:55
3. 7:35
4. 12:35
5. 3:27
6. 11:49

Page 130
1. 4:06
2. 11:11
3. 9:50
4. 1:22
5. 7:28
6. 7:48

Page 131
1. 11:40
2. 1:08
3. 11:03
4. 7:46
5. 342
6. 5:12

Page 132
1. 11:13
2. 1:04
3. 3:33
4. 9:06
5. 7:48
6. 1:52

Page 133
1. 2:43
2. 6:01
3. 12:15
4. 11:58
5. 11:32
6. 3:27

Page 134
1. 6:16
2. 3:47
3. 1:03
4. 6:44
5. 9:34
6. 6:32

Page 135
1. 10:26
2. 2:53
3. 10:50
4. 4:58
5. 11:08
6. 9:02

Page 135
1. 3:52
2. 12:22
3. 12:44
4. 12:49
5. 8:04
6. 5:04

Page 154
1. 2 hours
2. 4 hours

Page 155
1. 2 hours
2. 3 hours

Page 156
1. 1 hour 30 minutes
2. 3 hours 30 minutes

Page 157
1. 2 hours 15 minutes
2. 1 hour 30 minutes

Page 158
1. 1 hour 30 minutes
2. 45 minutes

Page 159
1. 1 hour 25 minutes
2. 25 minutes

Page 160
1. 1 hour 10 minutes
2. 2 hours 50 minutes

Page 161
1. 1 hour 10 minutes
2. 45 minutes

Page 162
1. 2 hours 30 minutes
2. 1 hour 25 minutes

Page 163
1. 1 hour 40 minutes
2. 2 hours 40 minutes

Page 164
1. 49 minutes
2. 47 minutes

Page 165
1. 3 hours 45 minutes
2. 2 hours 29 minutes

Page 166
1. 3 hours 18 minutes
2. 1 hour 56 minutes

Page 167
1. 2 hours 50 minutes
2. 2 hours 31 minutes

Page 168
1. 3 hours 44 minutes
2. 31 minutes

Page 171
1. 6:30
2. 1:35

Page 172
1. 3:15
2. 7:45

Page 173
1. 12:00 p.m.
2. 5:30

Page 174
1. 3:00
2. 2:15

Page 175
1. 5:45
2. 4:50

Page 176
1. 4:50
2. 45 minutes

Page 177
1. 90 minutes
2. 11:25

Page 178
1. 8:35
2. 1 hour 45 minutes

Page 179
1. 8:35
2. 3:10

Page 180
1. 1 hour 42 minutes
2. 1:02

Page 181
1. 5:35
2. 5:34

Page 182
1. 10:47
2. 12:43

Made in United States
North Haven, CT
05 November 2026

21772051R00104